MAPPING INFORMATION
The Graphic Display of Quantitative Information

HOWARD T. FISHER

MAPPING INFORMATION

The Graphic Display of Quantitative Information

Abt Books

Cambridge, Massachusetts

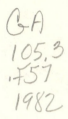

Reissued by arrangement with
University Press of America, Inc.
4720 Boston Way
Lanham, MD 20706

Library of Congress Cataloging in Publication Data

Fisher, Howard T.
  Mapping information.

  Bibliography: p.
  Includes index.
  1. Cartography. I. Cohen, Jacqueline Anna,
1944–   . II. Title.
GA105.3.F57   526   82–6858
ISBN 0–89011–571–0   AACR2

Contents

Tables

Figures

Foreword

When Howard Fisher suddenly passed away in 1979, he was in the process of writing a major work on thematic map design. He had planned to circulate some ten analytic papers, followed by a dozen demonstration sets that applied a wide variety of design techniques to sample areas, both real and imaginary. He intended to get comments and criticisms on the papers in their preliminary form and then use the revised versions as raw material in the drafting of his book. The first three papers had been issued before his death; the rest were in various stages of completion, as were the demonstration materials.

Howard's family and friends realized that the book he had intended to be the culmination of his career should be made available to others. His work on the organization and graphic representation of knowledge for which geographical location and spatial variation are relevant had already been recognized as original, creative, important—and frequently controversial— but few realized the full sweep of his ideas. He was not shy about challenging the sacred cows of traditional cartography, much to the discomfort of many who preferred tribal incantations to the systematic principles of map symbolism that Howard felt to be essential. As the creator of SYMAP—the first and still most widely used computer mapping program—he sought ways to forestall abuses of thematic cartography which the use of computers seemed likely to encourage.

With perhaps 75 percent of the intellectual effort for Howard Fisher's book completed at the time of his death, the services of Jacqueline Cohen, a technical editor, were enlisted to prepare his work for publication. Her skills were supplemented by those of several of Howard's former associates, principally Eliza McClennen. Their task was not to complete Howard's writings as he might have, but to accept Howard's general structure, as well as what he had written, and to select, reduce, and generally organize the voluminous materials left by Howard into the book that he had been developing. *Mapping Information* is the result.

The work is organized as a logical progression, from the most basic to the complex, to help the cartographically inexperienced reader learn. Byways are explored to help the reader think more clearly about graphically effective map design. The chaotic nature of classical thematic cartography is discussed, and fundamental principles are proposed. There is an attempt throughout to reveal how little is known, as well as to suggest guidelines for current practice. Those of us who worked with Howard recognize that each of these features reflects an element of his personality: a warm and supportive teacher; a sensitive and creative designer; a vigorous and demanding professional; and an inquiring researcher probing the unknown. We are richer for having known him. To his qualities this book bears witness.

Brian J. L. Berry
School of Urban and Public Affairs, Carnegie-Mellon University
Pittsburgh, Pennsylvania

Preface

This book is an attempt to deal with the issues involved in thematic cartography through a logical progression gradually increasing in complexity. It necessarily reflects my personal opinions, though I have tried to include those of others as well. The material has been arranged and the text composed primarily for the cartographically inexperienced reader. A serious effort has been made, however, to achieve both completeness and accuracy.

A second goal was to explore those bypaths that contribute to achieving the most graphically effective results. But each bypath led to others no less interesting. I determined to follow these paths individually, even when they merged with others for stretches that inevitably led to repetition. Such repetition appeared to offer important advantages. By probing and traversing the terrain from many angles and along many routes, a better understanding seemed possible. Given this approach, it should not be surprising if the book has turned into a hybrid, comprising elements of a textbook, a manual, and an encyclopedia.

Those unfamiliar with the mapping process might imagine that the spatial and quantitative facts are always given in final form, and the problem is merely to select suitable symbolism and execute the map accordingly. This is a great oversimplification, however, since these facts may be better represented after being manipulated in any of several ways. Spatial information may be modified through aggregation or interpolation, for example. Quantitative information may likewise be modified through aggregation into classes of equal or unequal size.

The basic premises expounded in this text apply to most thematic mapping problems and should serve as a solid foundation on which to build in more complex situations. Map design is illustrated by applying various procedures to a limited number of problems, producing alternative solutions. The virtues and limitations of different procedures can thus be meaningfully compared. Issues evoked by these mapping problems are discussed as encountered, even when they lead to consideration of more advanced concepts.

In recent years, much cartographic research has tended, not surprisingly, to deal with fragments of the total process. Yet, without an adequate intellectual framework, continuing progress in basic theory may be seriously handicapped. In view of the rapid acceleration of research and applications, significant amplification of the ideas in this book is expected. It is hoped that exceptions, however, will be rare.

The importance of employing basic terminology clearly and differentiating consistently can hardly be overemphasized. Many difficulties that continue to plague practical and theoretical work undoubtedly stem from semantic confusion. Similarly, certain traditional definitions have doubtful validity and should be avoided. New terminology has been employed here with care, however, and only when the change seemed vital to clarify thought.

As is well recognized, the issues involved in thematic cartography can be exceedingly complex, particularly because of the myriad possible relationships among situations, procedures, and symbolisms. It is hoped that this book may assist in sorting out those relationships, not only for difficult assignments but for more typical problems as well. Perhaps the chaotic terrain of thematic mapping will turn out to be an integrated and fascinating unity of diverse elements—all rational, interrelated, and mutually supporting.

Howard T. Fisher
Cambridge, Massachusetts

Biographical Notes

Howard Fisher was a designer by profession, and his concern in this book is the design aspect of thematic cartography. As his earlier professional work included neither geography nor cartography, readers may be interested in a brief resume of his background.

Born in Chicago in 1903, Fisher studied art history and architecture at Harvard, receiving a B.S. degree *magna cum laude* in 1926. Incidental to a distinguished career in architecture, which included extensive research and developmental work (in recognition of which he was made a Fellow of the American Institute of Architects in 1974), he became increasingly involved in city planning activities—and thus in the extensive use of maps.

As an outgrowth of education work undertaken for the United Nations, in 1957 he was invited to join the faculty of Northwestern University's Technological Institute. There he taught, among other subjects, the terminal design problem in civil engineering and a course in creative problem solving as a general process. It was while at Northwestern that he developed, in 1963, the SYMAP computer mapping system, originally conceived as an aid to consulting work being carried out for the City of Chicago's Planning Department. During the following year and a half his efforts were devoted primarily to effecting improvements in that program, to teaching its use, and to exploratory work with interested geographers, cartographers, and planners.

In February 1965 Fisher joined the faculty of the Department of City and Regional Planning in the Graduate School of Design at Harvard University. Shortly thereafter he founded the Laboratory for Computer Graphics, for which he received a grant of some $308,000 from the Ford Foundation the following autumn. Over the ensuing three years his efforts were devoted primarily to administrative duties, to launching the work of the Laboratory generally, and to assembling its staff of geographers, cartographers, and computer specialists. During this period, however, Fisher also worked on the refinement of programs, taught courses in Harvard's Faculty of Arts and Sciences as well as in the Graduate School of Design, and organized a number of national conferences.

Retiring from active administrative duties in 1968, Fisher was appointed Associate Director of the Laboratory and later Research Professor of Cartography. From that time forward, he worked exclusively on theoretical studies in relation to thematic cartography, both produced by hand and by computer.

During the summer of 1969, following a series of lectures given in Spain for the U.S. Department of State, Fisher devoted himself to research in the Map Room of the British Museum, to the organization of a major conference on computer mapping for the British government, and to visits with members of the geography or cartography departments of several British universities.

With further financial aid from the Ford Foundation and others, and with the assistance of a small technical staff operating within the Laboratory, he then began the research and writings that engaged him until his death in 1979, and of which work this book is the result.

Throughout his professional career, Fisher most enjoyed the challenges of design work and research activity directed toward design improvement. During the period of his initial cartographic work at Northwestern University, however, and later while serving as the Director of the Laboratory, he found but limited opportunity for library research of a theoretical nature. In fact, in contrast to the urgent need for problem solving at the level of practical accomplishment, theory as such probably seldom entered his mind—except in connection with the difficult issue of interpolation. Later, when he had the opportunity for more leisurely library research, he encountered difficulties in attempting to apply what he found in books and journal articles to the problems that he had under study. There appeared to be basic differences between traditional assumptions and what he found necessary for progress. The discussions in this book are in large part the result of his efforts to reconcile those differences, and to establish a sound theoretical foundation upon which future work might be based.

Allan H. Schmidt
Executive Director, Laboratory for Computer Graphics
  and Spatial Analysis

Acknowledgments

During the years devoted to his research in cartography, Howard T. Fisher sought stimulation and criticism from many people who shared his interest in mapping. He was always free in crediting the contributions of others to his work. In assembling the names that follow, the editors have tried to adhere to his example, and they extend their apologies to those whose names are inadvertently not recorded here.

Much of the form of this book is owed to the expertise of editor Jacqueline Cohen who, ironically, never had the opportunity to work with Howard Fisher. She undertook the enormous task of cataloguing and organizing ten years' worth of papers that ranged from published monographs to scant notes. Gaps in the material prevented her from following precisely the outline for the book that Fisher had projected. She did, however, use the maximum amount of available text and illustrative materials consistent with a reasonable degree of completeness to express Fisher's ideas in his own words.

In this task she was assisted by cartographer Eliza McClennen. Having prepared much of the artwork that appears in this book, she was able to coordinate the figures with the text and assist Cohen in understanding Fisher's methods and objectives. With cartographer Herb Heidt, McClennen worked full time with Fisher for four years, starting in 1970. Their continued close involvement with the project has been of great assistance in preparing the manuscript for publication and seeing the book through to publication.

At varying times, their fellow staff members were Jonathan Corson-Rikert, Alan H. Fisher, Ronald Gore, Elizabeth Durfee Hengen, Sis Hight, Ann Molineux, Mary Raymer, and Carolyn Weiss. Along with the numerous staff and participants in Fisher's 1970 Special Program in Thematic Mapping, they were all involved in the research and cartographic work included in this publication. The typing of manuscripts was carried out by a staff that included Betty Barnes, Nan Dealy, Julianne De Vere, and Lucy Richardson.

While pursing his research, Fisher kept abreast of developments in the rapidly evolving fields of computer graphics and automated cartography through his continued close association with the Harvard Laboratory for Computer Graphics and Spatial Analysis, of which he was the founder and first director. Allan H. Schmidt, Executive Director, and former Director Brian J. L. Berry provided much support. Staff members, past and present, including Nicholas Chrisman, Geoffrey Dutton, James Dougenik, William Nisen, David Sheehan, and Donald Shepard, all provided invaluable aid.

Local assistance also came from the staff of the Harvard Map Library and curator Dr. Frank E. Trout. Other helpful members of the Harvard community were Roger Fisher, Richard Land, William Mercer, Frederick Mosteller, and former Dean of the Graduate School of Design, Maurice D. Kilbridge.

Fisher enjoyed lengthy correspondence and discussions with many academics in the field of cartography. Among these were Dr. Kang-tsung Chang of the University of North Dakota, Dr. David Cuff of Temple University, Dr. George McCleary, Jr., of the University of Kansas, and J. T. Coppock and D. R. Macgregor of the University of Edinburgh. Their interest was always greatly appreciated, as was that of Walter D. Fisher of Northwestern University.

Howard Fisher's research was supported by grants from the Ford Foundation and by anonymous donation. This help was most gratefully accepted and used.

The papers as Howard Fisher left them at his death faced an uncertain future. Much of this uncertainty was resolved by the vigorous initiative of his sister, Margaret Fisher. Her generosity supported the organizing and editing of the papers into a form capable of being considered for publication. That this book has now appeared is due in the greatest degree to her kind and timely act.

PART I

Introduction

CHAPTER 1

Beginning Spatial Analysis

WHAT IS A MAP?

A map is a spatial analogue: its purpose is the understanding, portrayal, and communication of information that varies in space. Many problems of the modern world resulting from rapidly increasing urbanization—war, crime, narcotics, unemployment—involve spatial variables. The first step in developing adequate solutions is a thorough knowledge of the facts that exist and when, where, and how they vary over time. Then these facts must be understood and communicated. Information capable of being mapped is almost unlimited in variety; so are the needs and objectives of map users.

Mapping is a convention. We constantly accept conventions of this kind. A photograph seen with a magnifying glass consists of nothing but tiny all-black and all-white areas. Looked at from a distance, however, these become the different shapes and tones of a picture.

Maps usually represent geographic or physical space, but they can also portray and communicate information that varies in imaginary or abstract space. For example, a map showing relationships among the three basic variables of color would represent what is usually referred to as *color space*, a wholly imaginary construction.

Real space is usually defined in terms of such parameters as the distance north and south, east and west, up and down. Imaginary space, however, may be defined by any useful measurable scale. For instance, the parameters might be hue, darkness, and intensity; or the collar size and sleeve length of men's shirts (Figure 1–1). Displays of imaginary space may be referred to as charts, graphs, or diagrams, but the mapping problems involved are essentially similar to those encountered in mapping real space.

In a number of respects, our subject matter will be presented from new and unorthodox viewpoints. Many of the approaches recommended in the past were dependent upon the subjective judgment of the mapmaker. These

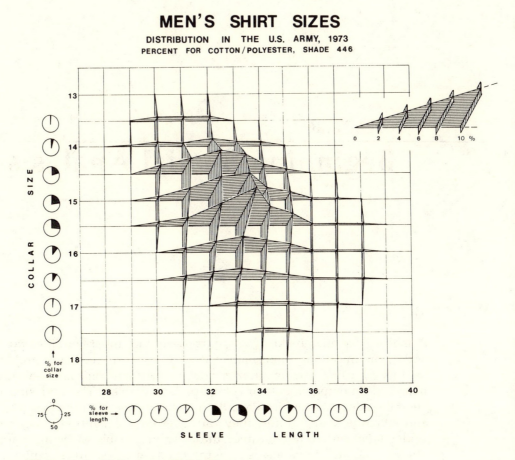

Figure 1–1   U.S. Army Shirt Sizes

approaches have now been standardized to a substantial degree, with a consequent increase in the objectivity and hence the credibility of maps. In addition, these recently developed methods provide a helpful flexibility in meeting the specific needs of individual assignments. Through a full discussion of alternative practices and symbolisms, the virtues and limitations of various methods will be observed and weighed. Informed choices can best be made when methods and procedures are fully understood, so that reliance on personal judgment—though never eliminated—is minimized.

Mathematical geographers, specialists in spatial analysis, statisticians, and others familiar with the traditional terminology should be reminded that this book is aimed at the production of maps by and for the educated public. We have chosen, therefore, to depart from established geographic terminology when traditional terms appear overly technical or less than precise.

Although the difference between mapping by hand and mapping by computer is minor on the conceptual level, the procedures described here will facilitate the use of the computer, and the reader is encouraged to use a computer in any instances in which it might prove advantageous. The same choices must be faced and the same decisions made when mapping by hand or by computer, but employing the remarkable powers of the computer requires a degree of precision that almost always proves helpful in other ways as well. In fact, the greatest value of the computer may be that it demands more rigorous thinking by the user.

Before you proceed in this book, quickly skim the chapter heads and illustrations to help avoid confusion. Almost every aspect of the subject is to some degree related to every other aspect. Thus, except at the introductory level, learning must take place in several areas simultaneously.

The historian can frequently proceed in a simple chronological sequence, the botanist through classification, the law professor in some logical sequence by subject matter. Most subjects can be taught in a more or less stepwise fashion. Mapping, however, involves the diverse aspects suggested by the words *spatial, quantitative,* and *symbolic.* Everything that exists must exist in some location, real or imaginary, and to some degree, known or estimated. And both aspects must be represented in a map through graphic symbolism. Thus even the simplest map—e.g., a kindergarten room drawn on the blackboard by a teacher—involves these three aspects simultaneously, according to its unique requirements. Since each of the three basic variables mentioned may involve a large number of alternative situations, the number of permutations and combinations can be considerable. Only a very few of these, however, will be rational and useful.

In a very real and fundamental sense, every map shows information that occurs in space. In this book, however, our concern will be less with the basic features typically portrayed in atlas maps than with more detailed facts regarding the nature of such features. Where the typical published map may show the physical locations of coastlines, we may be concerned with such variables as the emergence, height, steepness, and stability of bluffs; the character and extent of any tidal flats; the width and recreational value of beaches; the possible pollution of the beaches and adjacent waters; the number of shipwrecks; or other spatially variable information. Except in the simplest situations, descriptive words or data in tabular form cannot show spatial interrelationships.

In order to make a map, we must know the following three basic factors:
1. The overall study space in which we are interested, for example: the Milky Way, North America, Switzerland, London, the Wall Street financial district, a coal mine, an ant colony, the human brain, a tumor.
2. The information (or values) to be portrayed, for example: population or population density, barometric pressure, age, income, agricultural yields, angle of sunlight.
3. The locations to which the information applies, for example: stars, countries, counties, census tracts, city blocks, dwelling units, grid cells.

Whenever possible, the sponsor and mapmaker should consult together before this information is gathered, so that study space, values, and locations may be established in the way that is best for mapping. A successful map depends on the arrangement of graphic symbolism; it should be easy to comprehend.

The issues and procedures involved in making even the simplest map suggest a host of fascinating problems that characterize more representative work. Encountered in widely varying combinations, these problems pose numerous difficulties for the unwary, but for the student of cartography they present the opportunity for a great diversity of design solutions.

WHAT IS THEMATIC CARTOGRAPHY?

Thematic cartography has certain characteristics which distinguish it from general reference cartography. General reference maps place little emphasis on obviously quantitative differences (with the exception of terrain height). In contrast, thematic maps emphasize quantitative differences through the extensive use of graphic symbolism. General reference maps primarily depict physical characteristics on or near the surface of the earth and major political boundaries. Thematic maps may deal with any conceivable subject in any location, including nonphysical events and totally abstract or hypothetical matters. Individual thematic maps usually portray only one or a few subjects that are similar or clearly related (Figure 1–2). In contrast, many general reference maps treat a considerable number of subjects that are often un-related except for sharing a common study space (Figure 1–3).

Based on the preceding discussion we suggest the following definitions:

General Reference Cartography. The making of relatively standard-ized types of maps serving common or general reference purposes or users, in contrast to *Thematic Cartography* (q.v.). The following types of maps and any closely similar, when prepared in those forms most commonly encountered, would fall into this category: road maps, general reference atlas and wall maps, and topographic maps, including most aeronautical and nautical maps. All other maps fall into the thematic category.

Thematic Cartography. Mapmaking serving relatively specialized purposes or uses, in contrast to *General Reference Cartography.*

Maps thus defined as of general reference type are usually highly stan-dardized in format, periodically revised, and used over substantial periods of time. In contrast, maps of thematic type tend to be far less standard in format, less frequently revised, and of more transitory concern.

FUNCTION AND FORM IN MAP DESIGN

The beginning student in thematic map design—like the student of design in any field—must learn from direct experience, aided by a clear set of guiding principles. Good map design entails selecting the best combination of procedures and graphic elements for any given project. Circumstances will vary with the time, skill, equipment, and funding available. For example, designing a map for information study by a city planner requires a different approach than creating one for publication in an important city planning report.

Map purpose is fundamental to the overall cartographic process because it determines the specific information to be presented. Every map must have

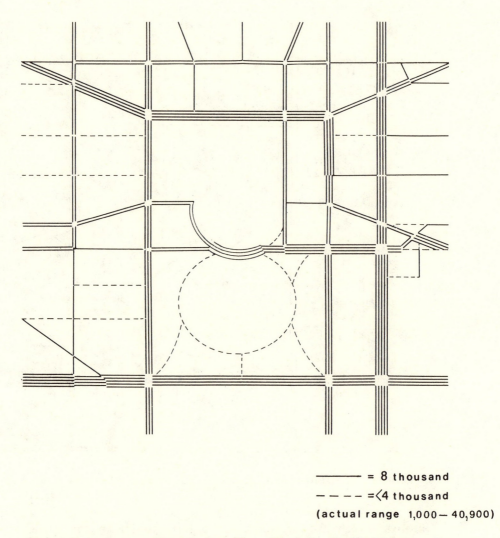

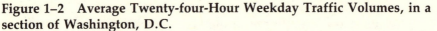

——————— = 8 thousand

— — — — =〈4 thousand

(actual range 1,000 – 40,900)

Figure 1–2 Average Twenty-four-Hour Weekday Traffic Volumes, in a section of Washington, D.C.

some purpose or objective, or time and effort would not be invested in its production. The objective may range from the trivial to the most serious, and it will determine the information to be presented, the audience that will be served, the time and money made available, and the final design. We will consider each of these items briefly except the influence of purpose upon design, which will be alluded to throughout and discussed in later sections.

The Information

The question of what information to communicate is the heart of the issue. Why present that information visually? The Census Bureau would say: "Because we believe there will be an interest in it." A commercial company would say: "Because we can make money from it." Others will have some

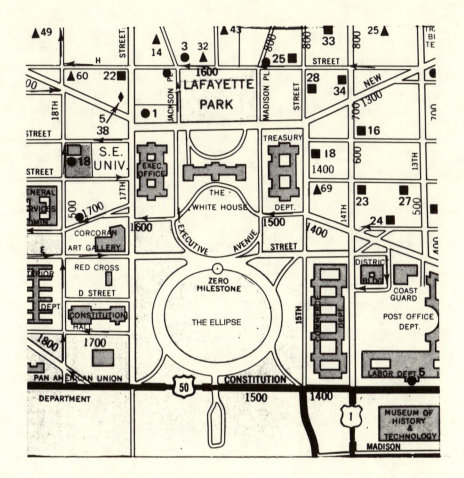

Figure 1–3   Section of Washington, D.C.

more specific reason: to educate, to inform, to illustrate an article or a book, and so on. Map design is usually a function of exactly what is to be communicated. When, for example, the purpose is to communicate information about the absolute population by counties within a study space, one must not show population per unit of area.

Choosing the specific subject to be presented is always the most crucial decision in any cartographic project and the starting point of the design process. The subject of interest may be Mexico City's subway system, but the specific subject will depend on the purpose of the map. Perhaps the purpose is merely to inform the public where the subway lines run and where the stations are located. Or the purpose may be to improve the scheduling of trains at peak hours of use, or to expand the system to serve a wider area. Obviously, the data for these three different purposes will be quite different. The special purpose is usually represented by the subjects chosen for portrayal. For example, the purpose of cadastral maps is land taxation, the subject is land ownership (name of owner and extent or boundaries of property), and possible uses besides land taxation include satisfaction of curiosity about one's neighbors or location of possible land sellers.

The Audience

The sponsor should have a fairly clear idea of the audience to which the map will be directed. For example, the primary audience for marine charts will be mariners and recreational boaters. The primary audience for a map showing deaths by cancer will presumably be persons concerned with its pathology or with the provision of treatment facilities.

Sometimes it may be risky to direct a map to a particular user or audience. For example, if a map is being made to persuade a prospective tenant to rent space in a shopping center, or to encourage a mayor to approve a proposal by the city planning department, it may be inadvisable to favor special interests or prejudices, since the map might be passed around for comment and advice or might even end up printed in the newspaper. (See also section on Ethical Questions, below.)

Occasionally, the intended audience may be so specific that the map may not be intelligible to the general public, such as a map for the blind that has braille notations. A more commonly encountered example is the case of meteorological maps, which require special symbolisms to be useful to their intended audience. Under such circumstances, lay persons who have reason to use such maps must familiarize themselves with the concepts and graphic conventions employed.

The language of the intended audience can be important if words are needed in the design. Under some circumstances, two or more different languages may be appropriate. Maps to be presented on television or to accompany oral presentations (slides or transparencies) should be as visually simple as possible. Due to the short viewing time, written matter should be minimal. And when the intended audience is unsophisticated or very young, maximum verbal and graphic simplicity is usually appropriate.

The size of the intended audience is also significant. A map intended to serve only a few people may, like a longhand note, be informal while still communicating the desired information. A map that will be more widely distributed demands greater care and perfection. The design of a printed map may be influenced by the particular mechanical procedures to be employed. Color should be used with great caution in a map that will be widely read by the general public, such as a display of evacuation routes, since many people are partially colorblind.

The degree of generalization employed is likely to be the design consideration most influenced by the intended audience. One should show the maximum detail that will not compete with or detract from the main message. In general, a well-designed map, like a well-written text, should be fully comprehensible to almost any interested person.

Time and Money

Obviously availability of both time and money will greatly influence design. Color printing takes more time than black and white, and is more expensive. Certain mechanical procedures may provide more flexibility but may require a high investment of time or money. If time is short or financial resources limited, a form of symbolism that can be designed and executed quickly should be chosen. People who design maps for newspapers are especially knowledgeable about ways to achieve acceptable results in the shortest possible time. Whether the final map is to be turned out by hand or by computer may also dictate quite different design solutions.

Ethical Questions

The usual objective of grouping data values into classes is to show the given information as meaningfully as possible, but classing (especially if not systematic) can easily be employed to hide significant differences or to give an impression of difference where none exists. A number of mapping procedures, valid when honestly employed, lend themselves to deliberate misrepresentation. For example, not revealing the number of locations, their sizes and shapes, or the extreme values present are ways of using maps for improper purposes. There is a genuine need for a small book entitled *How to Lie with Maps** to alert map users to deliberately slanted presentations. The same factors that make a totally honest display difficult make misrepresentation easy.

On occasion it may be legitimate to stress a particular viewpoint. To illustrate, if the subject is the intensity of proposed street lighting and the purpose of the map is to allay the worries of nearby homeowners who don't want unusually bright lights shining into their houses and gardens, one *might* symbolize the intensities differently than if one wanted to appeal to members of the local automobile club. Usually, however, any such slanting is not only unnecessary but undesirable, and the goal should be communication that is as effective and honest as possible.**

Consciously or unconsciously, a map designer exercises control through the choice of symbolism to which the scale of values must be oriented. Increasing darkness, for example, may be used to show either percent literacy or percent illiteracy in a population, depending on the purpose of the map. The choice of orientation is not necessarily a method of slanting of data, but may simply allow more effective value differentiation.

THE SPONSOR'S ROLE

An irony of mapping is that maps are necessary to understand the data, but you must understand the data to make a map. The way out of this chicken-and-egg situation is to make a few elementary assumptions, start, and then make improvements. Each successive map will increase understanding of the data, and increased understanding of the data will improve each successive map.

The sponsor and the mapmaker may be the same person but usually are not. In any case, we suggest considering their functions separately. The

*To be a companion volume to *How to Lie with Statistics* by Darrell Huff. New York: W. W. Norton & Co., 1954.

**From a different viewpoint, Cornelius Koeman comments in reviewing *Cartographic Design and Production* by J.S. Keates [*International Yearbook of Cartography* (15:190, 1975)] ". . . under the heading 'Map design' one would expect a specification: map design for schools; map design for tourists; etc."

sponsor should always try to provide the data to be mapped in its most complete and basic form, even if definite ideas have been formulated about how it should be processed. If the data is not yet gathered, the designer should be consulted in advance to suggest the parameters and degree of detail that will best serve the display.

The better the sponsor understands the mapmaker's needs, the easier it will be to furnish material in the form best suited to the mapping process. The better the mapmaker understands the sponsor's material and goals, the better suited the map will be to the subject matter. However, sponsors rarely understand all the ramifications of mapmaking and mapmakers are rarely as knowledgeable about their subject matter as are sponsors.

To the map designer, the number of subjects to be displayed is often crucial. Every thematic map has at least one subject of interest. The symbolism for this subject must often be displayed against a background of other information for the map to be meaningful.

Two or more subjects, related or unrelated, may be displayed on the same map. The subject that is used to make another subject on the same map more meaningful is the *background subject*. The distinction between a subject that is primary and one of supporting interest will prove useful, although it may not always be clear-cut. If the subject of a thematic map is winter wheat production, for example, other information may also be provided, such as the location of river courses, mountain ranges, railways, expressways, parks, and cemeteries. In such a map, these would not be considered additional map subjects. However, if another value (such as the acreage planted to wheat) is included solely in order to make the facts about wheat production more meaningful, it would be considered a background subject.

For even the simplest map, the sponsor will need to provide the variety of information suggested by the following topics, some of which were mentioned earlier and each of which will be discussed in detail later.

1. The subject matter or information to be displayed.
2. The overall study space to which the information relates. (If there are two or more areas of concern, it is best to consider each separately.)
3. The locations within the study space to which the information applies. Every map must have at least two locations.
4. The values to be displayed for each of those locations.
5. The applicable time. Every map must relate to one or more points in time or spans over time.
6. A variety of other items, such as the amplified title and circumstances of use, the preferred size of the map, whether it is to be reproduced, and if so, in what quantity and by what method (if decided).

The sponsor's ideas as to how the objectives may best be served should take the form of suggestions rather than instructions, so that the special knowledge, experience, and judgment of the mapmaker can be fully utilized.

CHAPTER 2
A Dialogue Between a Sponsor and a Designer

Designer: So the full title of the map you require will be "BIRDS OF SOUTH AMERICA." But what do you want to show about the birds of South America?
Sponsor: The different species that are found in South America.

D: But can't you just list them? Why do you require a map?
S: To show where they are.

D: Do you know where they are?
S: Well, more or less.

D: That sounds pretty vague—and therefore hard to map.
S: We know where they have been seen.

D: That sounds terribly specific—and equally hard to map. What information exactly do you have, and how is it organized?
S: For each species of bird, we have information derived from studies made within each country.

D: Is your information broken down within each country? If it is broken down, how is it broken down?
S: The information is given in terms of the major political subdivisions existing within each country. Incidentally, these vary considerably among the countries.

D: What kinds of information do you have that you want to show for each political subdivision of each country?
S: We know what species have been observed and something about the population of each species.

D: That word "something" suggests difficulty in mapping. Do you want to try to show the population of each species?
S: Oh, no! That information is only partly available, and even when available it is frequently too crude to be worth showing. All we want to map is what species are to be found in each location.

D: Just what do you mean by that? If one bird of a certain species has been seen once, is that species to be shown?
S: Yes—if confirmed by a responsible ornithologist—that is, by a member of the International Society of Ornithologists.

D: What about dates? What period of time are we concerned with?
S: Up through 1980.

D: Well, let's see now. Our study space is South America. Our locations are the political subdivisions of the various countries of South America. The value, as we would call it, to be represented for each location is, I assume, *presence*—with absence to be shown by no symbolism other than the white paper.
S: That is correct.

D: But now for a most crucial question: How many different subjects or species are there all together?
S: Oh, some 300.

D: Were you really thinking of showing 300 different symbols, one for each species? That sounds pretty ambitious, for both me and the map user.
S: No—I see that I should have stated that we only wish to map the major categories of species. Later we may wish to make a series of separate maps, one for each country, going into more detail.

D: So you don't want to map by species after all. For the first map, how many categories were you planning to deal with?
S: Not too many—around eight or ten. We haven't really made a final decision on that as yet. It may depend largely on how many can reasonably be shown on one map.

D: Well, we can probably meet your requirements, but bear in mind that there will have to be a different symbol for each category—and it may not be easy for the ordinary map user to remember eight or ten different symbols. In any case it may be all but impossible for anyone to be able to grasp the spatial extent of any one category over all of South America or over substantial portions of it. With eight or ten different symbols, it's all but impossible to concentrate on any one. Instead of making one big map for all eight or ten categories, what about a separate small map for each category? As well as being easier to understand, such a series of small maps would cost far less to produce, because each map would be very simple to make.
S: Why, that sounds like a fine idea—especially if they could all be small enough to be shown on one page so that comparisons among the different distributions could reasonably be made.

D: Well, with a number of small maps, it would certainly be easier to make comparisons. Let's talk about one such possible small map. What might be the best category to start with?
S: Let's assume it is waterfowl of all kinds, fresh-water as well as salt-water. Even salt-waterfowl, of course, are frequently seen on fresh water lakes and swamps.

D: But wouldn't the distribution of the salt-waterfowl extend considerably out into the ocean beyond South America? Where would you stop calling them South American birds? Perhaps a more appropriate title for your map would be something like "Waterfowl of the Western Hemisphere South of Panama."

S: Well, no. Let's stick to South America and not try to go beyond the newly proposed fishing limits.

D: Now, getting back to a possible title for the first of the series, would some working title such as the following perhaps serve?

BIRDS OF SOUTH AMERICA: WATERFOWL
As sighted by Members of the International Society
of Ornithologists through 1980
By major political subdivisions
of the 13 countries

S: That sounds fine. What you now propose would really tell anyone interested in seeing the map what they would find there.

D: There will thus be a series of small maps, each with a single specific subject—one of your categories. Of course, when several small maps are grouped together, a full separate title is not required for each. On each map, presence would be shown by a gray tone, in contrast to absence shown by the white paper. Color will not be required, and that will save a lot of money. Wherever the map is gray, we have a political subdivision within which the particular category being represented has been sighted by one of your responsible ornithologists. Wherever the map is left white, birds in that category have not been sighted. The gray tone will conform in its shape to the shape of the political subdivisions involved, so unless you wish to do so, you will not have to differentiate each subdivision by a separate outline. It should be possible to understand such a single-subject map at a glance, whereas showing all of your categories at once would probably have been extremely confusing, especially since your categories would, I assume, frequently overlap.

PART II

The Given Information

CHAPTER 3

Introduction to Cartographic Language

Editor's note: Most of the terms introduced here will be explained and discussed in detail in this part. The rest will be found in Creating the Display, Part III.

The terminology traditionally employed in thematic cartography tends to be complex, sketchy, and—in terms of modern developments—sometimes inadequate to essential distinctions. In such a rapidly developing field, some adjustment is inevitable.

The terminology used to describe mapping scale is sometimes confusing. Maps are usually much smaller than the actual spaces they portray, but they may be the same size or larger. For example, while a map showing astronomical information would be far smaller than the study space, a map portraying microscopic information would be larger than the study space. When relatively small areas such as cities are portrayed, the scale is thought of as large; with relatively large areas such as continents, the scale is thought of as small. A small-scale map embraces a lot of territory and a large-scale map, little territory.

In the phrase *map design process*, we refer to that aspect of cartography suggested by the common meaning of design, i.e., the process of deciding how the given information can best be displayed for maximum comprehension. The design phase is quite distinct from both the data assembly phase and the map production phase (by hand or by computer). When we mean *design*, we should not use the traditional term *compilation*, which tends to obscure the distinction between the assembly of the given information prior to the design process and that process itself. We exclude from design any preprocessing performed on the given information. We include within design any manipulation of the given information that serves design objectives (such as classing, the procedure of grouping given values of adjacent magnitude into a common category).

The common word *subject* refers to what is dealt with in any given thematic map. The words *component* or *factor*, which have sometimes been employed, are unsuitable, as they are used in discussing derived values.

We prefer the term *study space* to *study area*, as the latter ignores reality and neglects the important distinction between what is given and how it is portrayed.

As a general term for the areas to which the information applies, the word *location* is broadly applicable—regardless of the size, shape, or character of an area in the real world or as assigned on a map (see Chapter 6). Other terms pose problems. *Mapping unit* ignores the fact that locations usually exist regardless of whether or not they are portrayed on a map. The term *enumeration unit* is too restrictive for general use, as much cartographic information does not involve enumeration. *Statistical unit* is inappropriate when the information to be portrayed is not statistical. The term *unit area* has the same disadvantages as *study area*. The words *position*, *district*, *zone*, and *region* have special connotations that make them undesirable.

The traditional practice of applying terms signifying locational relationships to symbolism or values is confusing. For example, the word *point* is most often used in its geometric sense, which refers to a nondimensional location (to which a value may be assigned). If *point* is defined as nondimensional, it cannot be used to describe a symbol: a graphic symbol that is nondimensional would be invisible. Values may be assigned *to* points in order to use one of the three basic types of symbolism. But circles, squares, and other spot symbols are not *point* symbols, as there is no one-to-one relationship between symbolism type and location type. The term *point symbol* is therefore incongruous, meaningless, and misleading. We suggest *spot symbol* instead.

The word *value* appears to be the best possible term for quantitative information. Certainly *data* is not an alternative. Meaning literally "what is given," it glosses over the important distinction between given information and information derived from the given or introduced from outside sources later in the design process. One of the most common meanings of value, "Relative worth, utility, or importance . . . status in a scale of preferences,"* is especially applicable to ordinal scaling. A less common usage justifies *value* in connection with nominal scaling. According to one of the great English dictionaries of the past century, the word may also mean "import," defined as "the intrinsic meaning conveyed by anything," for which the synonyms are "sense, gist, tenor, substance."**

In relation to the classing of values, the term *span* is preferable to *interval*. Each class embraces a span (or stretch) of possible values between and including the class limits or extremes, rather than an interval (or empty space) between the extremes. We make the following suggestions regarding terminology for types of symbolism:

1. Instead of *point symbolism*, use *spot symbolism* as discussed earlier.

Webster's Third New International Dictionary. Springfield, Mass.: G. & C. Merriam, 1966.

**The Century Dictionary* (vol. 6, p. 6691, and vol. 3, p. 3013).

2. Instead of *line symbolism,* use *band symbolism* to stress its two-dimensionality.
3. Instead of *area symbolism,* use *field symbolism* to suggest its usually variable shape and to differentiate it from the two preceding symbolisms which also occupy area.

We suggest the following terminology for the quantitative analogues of symbolism:

1. The use of *count* rather than *countability,* largely for brevity.
2. The use of *darkness* rather than *tone,* to stress the recommended progression from light to dark, so that the greater the darkness, the greater the value represented.
3. The use of *extent* rather than *size,* to suggest reference to a specific aspect of magnitude—such as height or width—in contrast to size in general.

The distinctions among the different types of locations, values, and symbolisms are independent of the varying interrelationships that may exist in any given set of circumstances. Hence it is desirable to keep the terminology for each entirely separate.

CHAPTER **4**

Base Maps

Most thematic maps will be based at least in part on other maps. The most important considerations in choosing a source map are that the scale be appropriate and the necessary locations easily identifiable. At times, use of a copyrighted map, requiring the approval of the copyright owner, may be warranted. Material that is in the public domain and hence not subject to copyright protection is included on many maps containing copyright notices. Be cautious, however, when using copyrighted material of any kind.

Outline maps are common bases for thematic maps, but a wide variety of general reference maps may also be useful. While road maps sometimes serve as source maps, general reference atlas and wall maps are used more often. Aeronautical maps and nautical maps are also used.

The ideal base map, as well as being somewhat larger than the final map to be produced, should show little or no information in addition to that necessary to produce the thematic map. Usually the only information desired is the locations to which the values of interest will apply. The best source map is thus the simplest map available of suitable scale that shows the geographical, political, or statistical divisions of interest. If the final map also requires certain background information—such as facts regarding major cities, coastlines, railways, or rivers—this information should preferably be obtainable from the source map. However, the use of computers facilitates the assembly of information from two or more different maps, including the resolution of differences in scale or map projection.*

*A map projection is the transformation of coordinates measured on the earth's surface (spherical coordinates, or latitude and longitude) to coordinates measured on a flat surface (planar, or x,y coordinates). When selecting a source map, one should always bear in mind that the particular map projection to be employed may prove of considerable importance. This is especially true if the study space covers any substantial percentage of the earth's surface. (The reader is referred to Robinson and Sale, among others, in the Bibliography.)

United States Geological Survey quadrangle maps may be used as source maps. U.S.G.S. sheets deal with more than a hundred subjects, including such exotica as windmills and sunken wrecks. (The agency issues a "Topographic Map Information and Symbols" listing.) Although there are many subjects for which topographic maps are not useful, they are excellent sources for both hand-drawn and computer maps based on political boundaries or certain other types of locations (such as rivers, canals, and roads).

In computer language, a cartographic data base (CDB) contains the x,y coordinates and identifying code which define the locations of a study space. CDBs are usually created using a selected map projection, though the coordinates may be expressed as latitude and longitude only. Like an outline map for hand-drawn work, a CDB may be used repeatedly for thematic maps of different subjects that are produced by computer. U.S.G.S. sheets are now available as CDBs. The Central Intelligence Agency distributes the World Data Bank-1 CDB, and the U.S. Census Bureau has the Urban Atlas series—a standard source for large-scale mapping of urban areas—in both map sheet and CDB form.

Occasionally, no suitable source map will exist for a desired thematic map. For example, if you are mapping the species of individual trees existing within a newly established arboretum, there would probably be no source except a survey or outline of the study space as a whole. Under such circumstances, the applicable locations (those of the individual trees) would have to be established for the first time. A study of a new archaeological site, a micro-climatological investigation, or research on the spread of disease through human tissue would also involve establishing new locations or an entirely new study space. Such situations are most likely to be encountered when the study space is small, as in the several examples given, but may exist with study spaces of any size—for example, in mapping climatic regions.

CHAPTER 5

"A Dialogue" Revisited: Subjects and Titling

DIFFERENTIATING SINGLE- AND MULTI-SUBJECT MAPS

A single-subject map displays information about one subject of specific concern, exclusive of the study space parameters or any supplementary subject matter. A map showing population density in 1980 for the states of the U.S.A., for example, would be thought of as single-subject, although additional subjects might be included incidentally. Single-subject maps will always be the easiest to understand.

Multi-subject maps display information about two or more subjects. The information may or may not apply to the same locations. A map showing each of ten kinds of metal production would be a ten-subject map. The subjects may be dissimilar or similar in character, but every map (including those showing change over time) must be considered multi-subject if two or more different values apply to the same location, as when comparing parallel subjects or disclosing interesting relationships among nonparallel subjects. The larger the number of subjects, the more difficult the map will be to comprehend, like a meeting in which several persons are talking at the same time.

A multi-subject map will be easier to understand if (1) the locations are the same for each subject (unless two or more contour-type maps are being combined); (2) the measurement units are identical for each subject; and (3) the various subjects are clearly differentiated.

A multi-subject map that has only one value per location—such as dominant crop, dominant language, dominant political party, or a dominant soil type—may be a presence-or-absence map or may show the degree of dominance, as 65 percent Democratic in one area and 57 percent Republican in another. A weather map displaying both temperature and rainfall would be a multi-subject map. Assuming for simplicity two levels of each subject, there would be four possible combinations: cool-dry, cool-moist, warm-dry, and warm-moist. Such a map would display the presence-and-absence only of each combination for the given locations.

Subjects that Change over Time

A multi-subject map will usually be easier to understand if the study time can be the same for each subject—unless the only difference is time. When change over time is involved, each instant or span of time for which there is a value constitutes a separate subject. For example, if the growth of a city's population is to be mapped and information is in ten-year intervals from 1930 to 1980, the map will be a six-subject map.

Change over time in terms of movement represents the same situation. The path of a ship at sea, the route of the Lewis and Clark expedition, or the trail of a fugitive from justice are based on knowledge of position at a series of specified times. This also applies to movements of population. Suppose you are interested in mapping the migration of a herd of elk or the movements of an army. You would determine and show the relative amount of time spent by the population in each data zone.

If a continuous trail is being traced, like the track of a skimobile passing over a snow surface, skid marks on a pavement, or the path of the sun, the map could in a sense be regarded as single-subject, communicating the facts within one span of time. Normally, however, such information—no matter how voluminous or closely spaced in time—is regarded as discontinuous, as in the separate frames of a motion picture.

CREATING EFFECTIVE TITLES

The map title should clearly identify the subjects of primary concern, so that the user can immediately appreciate what is being symbolized and what is being treated as *ground*. A map titled "Areas Subject to Flooding" would place visual emphasis on those areas; all other areas would be subordinated as the ground against which the flooding stands out.

A final title should give the information most essential to differentiating the particular map from all other maps. Titles should be as concise as possible. If the main title cannot be made definitive without becoming too long, a supplementary title may be employed. Fundamentally, the title should state the particular variables being portrayed.

The single most revealing question that a mapmaker can ask a potential client or map sponsor is, "What *exactly* is the full title that would identify this map properly in a bibliography?" The answer to this question will usually give only the vaguest impression of what is really desired, for example, "Birds of South America." Then a dialogue such as that in Chapter 2 may be necessary.

The nature of the locations to which the values relate should always be stated. If the locations vary for the different subjects in a multi-subject map, the locations applicable to each subject should be stated, as in "Votes by City Wards and Population by Census Tracts."

Units of measurement should usually be identified and differentiated by subject if necessary. (This will be especially important during our impending metric conversion.) When the values are dollars or other easily stated units, they may be identified in the value key. However, when the units employed are more complex or require a long explanation, they should usually be presented in the title. For example, if the map deals with the value of television sets manufactured in the Orient in 1978 expressed in dollars based on 1975 exchange rates, it would be best to state that once in the title rather than to squeeze it into the value key or a note.

Adequate dating information should always be included unless it is implicit. Giving the year will usually be sufficient, but more precise dating may be required. When a span of time is important, state it in the title. "Automotive Traffic Volume between 8:00 and 10:00 A.M., on Weekdays during July, 1984" actually includes three spans of time: the 8:00–10:00 A.M. span, the weekday span, and the month of July, 1984. In multi-subject maps it may be necessary to note different dating information for various subjects.

Good examples:

ILLINOIS
Licensed Pharmacists, 1980
By Counties

TENNESSEE VALLEY
FULL POOL RESERVOIR HEIGHTS
in feet and tenths

as officially established by
The Tennessee Valley Authority
1970

From what has already been stated, it is clear that good titles for multi-subject maps can become quite complicated in spite of every wish to keep them simple. Even the simplest will name two or more specific subjects that are subdivisions of a general one. For example:

CALIFORNIA CITRUS PRODUCTION, 1980
Oranges, Grapefruit, Lemons, Tangerines
By Counties

When the specific subjects are separate, the title becomes more difficult. For example:

CAMBRIDGE, MASS.
POPULATION (by blocks) & PUBLIC LIBRARY RESOURCES
(books by libraries)
1975

When several maps will be used in a series, it is important to create parallel titles for comparable subjects. For example, if the presence of heating and plumbing in single-family houses is to be viewed in separate maps, the map titles should be parallel, such as: "Percentage of Single-Family Houses with Heating" and "Percentage of Single-Family Houses with Plumbing."

Programmers of computer mapping systems, note: It is crucial to include a map title on each and every map run. Otherwise, the maps may fail to be identified later, particularly with a large run. Maps that seemed clear and unforgettable when run can quickly become maps of doubtful meaning—hence worthless. Information about data sources should also be included.

The following points should be remembered when creating titles:

1. Good titles can be very useful in a variety of ways.
2. Titles frequently will have to be somewhat lengthy. At the very least, they should reveal the subjects which are presented on the map.
3. It may also be appropriate to include units of measurement and time spans in the titles.

Catalogue Card Information

The Library of Congress specifies the format of catalogue cards, for which the mapmaker or map publisher furnishes information. It is customary to provide detail such as the size of the map, the scale of the map, the map projection used, and whether the map is in color or black and white. Frequently additional information will be desired, such as whether the map as issued is folded, and if so, the folded size; whether information sources are named; whether a *coverage* or *reliability diagram* is included; the nature of any background data; the orientation, if not north up; whether insets are provided; and so on. Sometimes even finer detail is included, such as the particular standard parallels shown, the place of issue if different from the official source, or the name of the printer.

Curiously, information of far greater value to potential users is usually omitted, such as the number of locations employed, or whether the values have been classed, and if so, the number of classes employed and the type of classing. It is not even customary to state the basic type of symbolism employed or whether interpolation has been used. Information of this kind is usually far more important for thematic maps than the particular map projection employed or the size and scale of the map.

CHAPTER **6**

The Study Space and Locations

DESCRIBING THE STUDY SPACE

The study space is the entire space of interest or concern within which the mapping information lies. The study space may exist in the physical world or may be conceptual, as previously stated. Actual parameters are those used in measuring physical distance, such as light-years, miles, kilometers, meters, feet, inches, centimeters, nanometers, or degrees (as in latitude and longitude). Conceptual parameters are those of any other character—dollars, tons, years, births, accidents, percentages, degrees celsius, or miles per hour.

A conceptual study space may be 3-dimensional, as is actual space, or 1-dimensional, 2-dimensional, or n-dimensional (four or more dimensions). Spatial variability cannot exist at a nondimensional point and therefore requires at least one dimension.

A 1-dimensional study space is wholly conceptual, and can only be described by a single parameter. If the parameter is time, each location may then be 0-dimensional (a point along the time span comprising the study space):

> 1900----1920-1925---1940--1950

or 1-dimensional (a time span along the time span comprising the study space):

> 1900----1920-25--1935-45-1950

A 2-dimensional study space is described by two parameters; for example, time and expense. A 3-dimensional study space may be wholly conceptual, represent "real" space, or both. Only three parameters can exist there. Given the parameters time, expense, and manpower, for example, each location within the study space may then be a point anywhere within the time-expense-manpower volume (0-dimensional); a time span, an expense span, or a manpower span (1-dimensional); or a time-expense subarea, a time-manpower subarea, or an expense-manpower subarea (2-dimensional).

26

An n-dimensional study space is wholly conceptual. Four or more parameters can exist in terms of the study space. If the four parameters are time, expense, manpower, and production, the last three might be treated as a 3-dimensional situation for each point or span in time that is considered.

A compact study space has a small surface compared to its volume (if 3-D), or a small perimeter compared to its area (if 2-D). A sphere is the most compact 3-D shape, a circle the most compact 2-D shape. Intermediate describes a 3-D space like a two-story motel or a 2-D space like the representation of Idaho. A spread-out space has a large surface compared to its volume (if 3-D), like a snake, or a large perimeter compared to its area (if 2-D), such as the representation of Chile. The outer limits are the surface of a 3-D study space or the perimeter of a 2-D space.

The locations within the space may all be in full contact with one another or with the outer limits and hence fill the study space (like the counties of Georgia); may have no contact with one another or with the outer limits (like the sampling stations used in an air pollution study). Where there is no contact, *interspace* exists—that area which, though within the study space, is not within any location. For example, in a map showing the Standard Metropolitan Statistical areas of the northeastern United States, all other regions are interspace (see Figure 8–3, p. 52).

ASSIGNMENT TO LOCATIONS

Locations are those positions within the study space to which the values to be displayed are assigned. The process of assignment, though not identified by any name in the literature, is basic. It means the transformation and representation of a study space or location by a point, a line, or an area on a map. Assignment to a point requires only a single set of coordinates; the point will not grow larger if the map is enlarged. Assignment to a line requires at least two sets of coordinates; the line will grow longer (if it is not straight, will spread over a greater area) but it will become no wider or higher if the map is enlarged. Assignment to an area requires at least three sets of coordinates; the area will grow larger if the map is enlarged.

There is always a direct relationship between each location and its associated value which is not affected by the assignment process. If the value 387 applies to a city (3-D), it will apply whether the city is assigned to a *base area* (2-D), a *base line* (1-D), or a *base center* (0-D). Assignment of a location necessarily involves corresponding assignment of the value that is related to it.

Single-layer study spaces are the most common situation encountered in practice. This is fortunate, since with such study spaces the value for each location can be assigned to the area of its base or to the center of its base, a highly desirable practice for graphic simplicity. Figure 9–2A (p. 61) shows the base areas of the locations in an imaginary study space we will call *Foursquare* (more about this in Chapter 9), with the applicable value for each location printed directly on it. Figure 9–2B shows the base centers of the locations, with the value applicable to each location printed nearby.

When base areas are elongated, such as for locations representing trans-

Questions

1. Give an example of a map based on one physical distance parameter and one parameter of another type.
2. If mapping from scratch, what is the crucial theoretical distinction between mapping the earth's surface above and below sea level?

Answers

1. Distance from the central police station; deaths from auto accidents.
2. Above sea level, the surface is visible, and, based on what is seen, measurements can be taken at significant selected locations. Photogrammetry can be used. Below sea level, the surface is normally hidden from view.

Editor's note: Some of the questions and answers in this book were taken from a course given at the Harvard Graduate School of Design in 1975 by Howard Fisher and Carolyn Weiss.

portation arteries, there is a third type of assignment that can prove useful. This is assignment to the base centerlines of the locations (Figure 6–1). Such centerlines, like the boundary lines of base areas and points, are actually devoid of width and height and thus invisible, although they are represented by ink lines. Centerlines conveniently suggest location position when the map scale would render area assignment graphically unsatisfactory.

Planners are concerned with data for locations such as states, counties, townships, census tracts, and blocks, as well as averages over these. If you are mapping the world, it's best to collect information by countries. If you are mapping countries, it is best to collect it by states, and so on. There is always a finer subdivision.

For most given data, the locations to be used are usually established first and the applicable values are then related. For example, in planning the collection of information relative to housing conditions, the U.S. Census Bureau first decided upon city blocks as the locations to be employed, and then gathered the data on housing conditions for each block.

Sometimes the process may be reversed. If the subject of interest is forests destroyed by fires, the categories might be simply "intact" and "destroyed." The locations would be established to agree with these classes. Classification is usually the first step when the values can be determined by visual observation or photographed, or when they are impressionistic and difficult to measure precisely, such as a map of military status with the categories "under our control," "control in question," and "under enemy control."

When locations are determined first, the entire study space may be of interest, or some portions may be ignored or considered to be secondary. For example, if the subject is dwelling units by blocks, the alleys between the blocks would probably appear as empty space. If, however, the subject is average July temperatures reported at major weather stations, the space between the weather stations would probably be symbolized through interpolation.

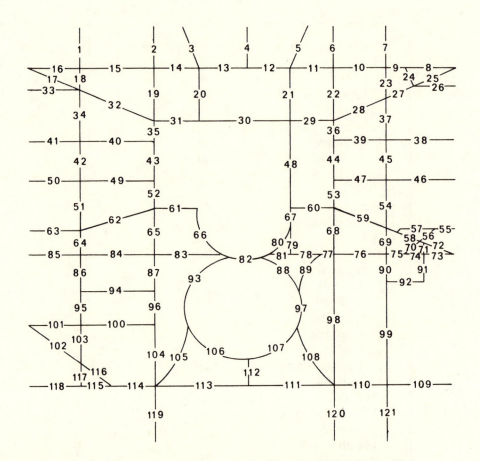

Figure 6–1 Locations from Figure 1–3 Assigned to Base Centerline

Natural, Administrative, and Display Locations

Locations may be natural, such as islands, oceans, rivers, watersheds, hills and dales, peaks and pales, forests, and pasture areas. They may be political or administrative units, such as countries, counties, school districts, and Federal Reserve areas. They may be statistical zones (census tracts, quarter tracts, and grids), or they may be established for purposes that are neither political nor statistical—canals, highways, man-made reservoirs, real estate parcels, cemeteries, and towns. Locations may also be established for display purposes only. We will see examples of this type of location in Chapter 11.

Generalization

With hand drafting, the locations employed may be traced over a base map or air photograph. For computer mapping, they must be digitized, i.e., described quantitatively in terms of latitude and longitude or an alternative grid system. In either case, the degree of detail selected may vary widely. Assuming the use of the computer and an irregular location, as few as five or six points might be used around the perimeter or as many as fifty. For most thematic maps, a considerable degree of generalization is not only acceptable but desirable.

Considerations in Treatment of Locations
TYPE
[　] Point
[　] Line
[　] Area
[　] Volume
　[　] One-layer
　[　] Multi-layer
　[　] Other
NUMBER
[　] Few
[　] Intermediate
[　] Many
SHAPE
[　] Similarity
　[　] Identical
　[　] Different
[　] Character
　[　] Compact
　[　] Intermediate
　[　] Spread-out
　[　] Mixed
EXTENT
[　] Roughly equal
[　] Diverse
[　] Highly unequal
CONTIGUITY
[　] Contiguous
[　] Separated
[　] Mixed
AS PERCENTAGE OF STUDY SPACE
[　] Total
[　] Major
[　] Intermediate
[　] Minor
DATA AVAILABLE
[　] By map
[　] By machine-readable tabulation
[　] By other

Following even minute irregularities around a location may greatly increase costs and attract the viewer needlessly to the location base outline rather than the focus of interest. However, when generalizing location outlines, it is important to recognize that even minor features should at times be included if they have significant recognition value. For example, when showing the Atlantic coast of the United States, it is customary to preserve Cape Cod, even though it is small in relation to the total shoreline.

If we flew in an airplane over a city with the objective of comprehending the spatial organization of the city as a whole, we would concentrate our attention upon major features rather than upon local details, no matter how interesting the latter might be. It is the classic problem of having to choose to see either the forest or the trees.

It will usually be necessary for the map designer to make a conscious choice between generalization and particularization. Each has its value, but if the main goal is good overall comprehension, we will usually need to generalize to a substantial degree. If our concern is with individual values alone, a table will usually serve and a map is unnecessary.

As we shall see shortly when we start to consider alternative symbolisms, many degrees of generalization and particularization are possible—depending not only upon the symbolism selected and the manner in which it is to be used, but also on the map size. No method has ever been invented which is capable of optimally serving both goals simultaneously.

Graphic symbolism cannot communicate quantities (except for small whole numbers) as precisely as Arabic numerals. Under some circumstances the latter can be used in combination with graphic symbolism, but unless the number of locations within the study space is quite small or the map unusually large, Arabic numerals will destroy the effectiveness of the graphic message.

CHAPTER 7
Values

Values are the information to be displayed, usually expressed in quantitative form. They may be positive, negative, or a combination of both. If there are two or more subjects, values must be available for each.

Values may vary in space:

—anything related to two or more spaces

—anything that varies in time, though related to a single space

—things that vary in space and time, such as population, altitude, or public opinion

Or not:

—temperature at a given moment at a particular location

—the decision handed down by a particular court in a law suit

—the height of a building

—the gross national product of a given country in a particular year

They may not vary in time or space, but may vary in conceptual space, such as the values represented by a chart of temperature and humidity.

Occasionally, a sponsor may need a map for which he has inadequate information, and the mapmaker may be asked to do the necessary research. Such activity is preparation for and not a part of the display process.*

Population density may be the subject of interest, for example. If the only values available are the population and area of each location, the desired density values can be easily computed. However, for a more complicated

*One good source of information for a wide variety of U.S. statistics is the *County and City Data Book,* published by the Census Bureau every five years. Breakdowns are given by the following locations: regions, states, standard federal administrative areas, standard metropolitan statistical areas, standard consolidated areas, state economic areas, economic sub-regions, counties, cities, congressional districts, central business districts, major retail centers, census tracts, city blocks, and minor civil divisions.

subject, such as a projection of population density, the sponsor should specify the procedure for determining those values.

While planning a map, you might discover that the values happen to be the same for every location. Would you proceed? You could, if you wanted to visualize where the locations were. Or, if you were making two or more maps of different subjects for the same study space or of the same subject for different study spaces, a map showing the same values throughout might be useful.

ADDABLE AND NONADDABLE VALUES

For mapmaking, there are two fundamentally different types of values. It is essential to differentiate between them, as under some circumstances mapping methods suitable for one type may not be suitable for the other.

Fortunately, a simple test distinguishes between the two types of values. Imagine that we desire to merge two or more adjacent locations into a single location, as in going from mapping data by counties to mapping data by states. Addable values may be summed without further manipulation of any kind. Nonaddable values will have to be manipulated in some other way. To illustrate, let us assume that the two adjacent locations shown below are to be merged into a single location.

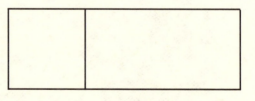

value = 15   value = 20

The original values are shown. What value applies to the combined location? If the values represent number of persons, they are of the addable type, and the new value is 35. If they represent number of persons per square mile, they are of the nonaddable type. The new value will have to be derived by other means, such as interpolation (which will be discussed later in this chapter).

Locations should usually be differentiated when addable values are employed, though with certain symbolisms, stating the number of locations may be all that is required. Traditional dot maps use addable values.

Addable values are usually simple counts of objects, events, or even money, as in the number of detached houses, the number of cases of measles, or the number of bankruptcies. Mapped, they would bear such titles as "Detached Houses by States in the Coterminous United States," "Cases of Measles by School Districts in Hampshire and Franklin Counties," or "Bankruptcies by Counties in New England" ("by" indicating location).

Such values are also called absolute, and may be established in a variety of ways: by direct observation or inquiry, as in much census data; through the study of aerial photographs or other maps; by the use of instrumentation; or by some combination of procedures. The value established for any location should apply to the entire location to the extent defined by the map title. "Urban Population by Counties in New Jersey" means all persons residing in urban areas in all counties in New Jersey, regardless of how "urban areas" are defined or whether or not they completely fill each county.

Values that cannot be established for an entire location, such as land elevation or temperature, are derived values. They are always an average of some kind and will usually be nonaddable. Rates are in this category, including density and the various forms of percentage—per 100, per mil, and p.p.m. (parts per million). With birth rates, the phrase "per 1000 persons" is usually understood (Figure 7–1); with average adult height, "per person"; and with temperature, "per sampling point."

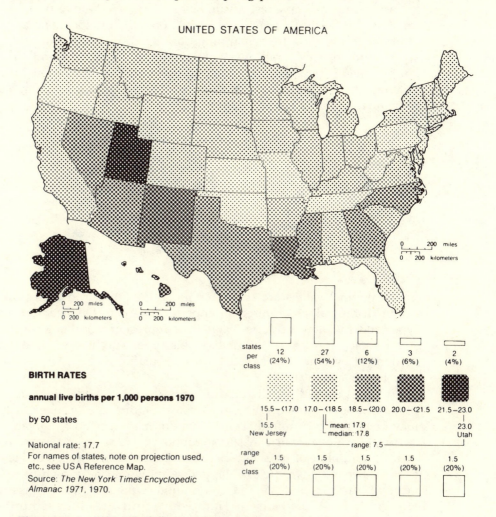

Figure 7–1 U.S. Birth Rates, Derived Values

Maps using derived values do not require the number of locations to be specified. Titles such as "Fleas per Monkey by Zoo" and "Average Value of Crops per Harvested Acre by Farm" demonstrate that our only interest is in the rate of something per location. However, the procedure used to determine derived values should be revealed, whether arithmetic or by some other means, such as use of a hygrometer to establish relative humidity.

The distinction between absolute and derived values can frequently be useful, as illustrated in relation to the wording of map titles, but in terms of map symbolism it is not always definitive. For example, the figures for a population projection to the year 2000 (if they are not mere guesses) would certainly be derived values, yet, since they are addable, they could be symbolized in exactly the same way as absolute addable values. In contrast, the distinction between addable and nonaddable values is definitive and may be used with confidence to determine the appropriate symbolism in any given situation.

TEST: If values for different locations can be added meaningfully, they are absolutes—and if they are to be displayed as absolutes, you need to differentiate locations. If they are to be displayed as rates, differentiation is not required. Conversely, when locations are not differentiated, the map can only be read as a rate map.

1. Assume that darkness symbolism designates 10 to <20 persons per square mile. Since the designation is "per square mile," the specified value span of 10 to <20 must apply everywhere the tone appears, without regard to particular locations. Hence locations do not need to be differentiated.
2. Assume that darkness symbolism designates 10 to 20% of the population over the age of 65. Since the designation is the number of persons over 65 per 100 inhabitants, and since it makes no difference what locations are inhabited, there is no need for the locations to be differentiated. Showing location outlines might be useful for general background purposes, but would in no way contribute to a better understanding of the spatial distribution of the values within an area of darkness tone. (It would be useful, however, to know the population of each location.)
3. Assume that darkness symbolism designates an average of 10 to <20 feet elevation of the ground. Since the designated quantity applies equally to every location within the tone, there is no need to differentiate the location.

There is another, more traditional method of categorizing values. This system, used by statisticians, divides the quantitative information to be displayed into four types: nominal, ordinal, interval, and ratio.

NOMINAL AND ORDINAL VALUES

Nominal values are names for things which are not rankable, having no sequential significance. A map showing cows, horses, and sheep would be of the nominal type.

Ordinal values can be ranked and have sequential significance, such as "small, medium, and large." The distinction between nominal and ordinal values is often arbitrary. Comparable nominal values can often be grouped as different levels of the same subject. The progression "squabs, pigeons, chickens, geese, and turkeys," for instance, can be viewed as birds increasing in size or as five separate subjects or values. If the subjects are rain, sleet, snow, and hail, they can be thought of as precipitation at four levels of increasing density, and therefore ordinal values. Built-up areas, farmland, and woodland can be ranked by decreasing intensity of use. Residential, commercial, and government land, however, cannot be ranked in a reasonable sequence.

Presence and Absence

The distinction between nominal and ordinal is further blurred by the concept of presence and absence. Nominal values—straightforward, independent entities—are either present or not present. If you are mapping agricultural land by crops—sweet corn, oats, and wheat, for example—the presence of sweet corn only in field 20 means the absence of oats and wheat in that field. In listing the data for field 20, the entry would be "Sweet corn yes, oats no, wheat no." In a sense, each subject is being ranked in two levels: At any particular location, it may be present to a significant degree or not (absent). In order to show presence, absence must also be considered, even in a single-subject map.

Imagine a study space composed of ten locations, with houses built on seven locations and the other three vacant. Assume that you wish to draw an informal map showing improved parcels. You have a supply of pens, brushes, ink, and paints, and a sheet of clear plastic to draw on. How would you make a neat final map? You would have to do two things: (1) show the presence of improved land by some positive means, such as a colored tone spread over those locations with houses; and (2) show the absence of improved land by some means, such as a white tone spread over those locations without houses. For a map that can be read without confusion, both steps are essential. Presence and absence are as inextricably related as the two sides of a coin.

To show presence for two or more related subjects, a different symbolism would have to be used for each; these must be chosen so that one symbolism does not hide another. A single type of symbolism could be used for absence applied to all of the subjects. Assume a study space composed of twelve locations, three improved with houses, two improved with nonresidential structures, four improved with both houses and nonresidential structures, and three which are vacant. The presence of improved land might be shown by black lines drawn horizontally to show the presence of houses and vertically to show the presence of nonresidential structures. The four locations

Assume an experimental agricultural farm with 20 square test plots laid out as indicated below, each plot one acre in area and numbered for identification as shown.

1	2	3	4	5
6	7	8	9	10
11	12	13	14	15
16	17	18	19	20

Each plot may or may not be planted to one or more of these five crops: corn, wheat, barley, oats, pumpkins.

The given information as to what is actually planted in each plot is shown in the following table:

PLOT	Corn	Wheat	CROP Barley	Oats	Pumpkins
1	X	O	O	O	O
2	X	O	O	O	O
3	O	O	X	O	O
4	X	O	O	O	O
5	O	X	O	O	O
6	X	O	O	O	O
7	X	O	O	O	X
8	O	O	O	O	O
9	O	X	O	O	O
10	O	X	O	O	O
11	O	O	O	O	X
12	X	O	O	X	O
13	X	O	O	X	O
14	X	O	O	X	O
15	O	O	X	O	O
16	O	O	X	O	O
17	X	O	O	O	X
18	O	X	O	O	O
19	O	O	O	O	O
20	X	O	O	O	O

X = present
O = absent

The facts may be summarized in words as follows:

In Plots #1, 2, 4, 6, and 20 only corn is present, the other four crops being absent.

In Plots #3, 15, and 16 only barley is present, the other four crops being absent.

In Plots #12, 13, and 14 corn and oats are present, the other three crops being absent.

In Plots #5, 9, 10, and 18 only wheat is present, the other four crops being absent.

In Plots #7 and 17 corn and pumpkins are both present, the other three crops being absent.

In Plots #8 and 19 none of the five crops is present.

Instructions

1. Tracing over the given map, including its border, make a rough freehand one-subject map showing where corn is present. (See Figure 7–2A.)

2. Tracing over the given map, including its border, make a rough freehand five-subject map showing where each of the five crops is present. (See Figure 7–2B.)

Which map was more difficult to create symbolism for? For a sound solution to this problem, is it necessary to differentiate each of the various locations?

improved with both types of structure would thus be symbolized by both horizontal and vertical lines. The unimproved parcels would be symbolized by white tone only.

Interval and Ratio Scaling

Interval and ratio scaling both involve continuous variables. Values of the interval type are expressed in numbers on a scale that is measured from an arbitrary starting point, not necessarily true zero. The same unit of measurement must be used throughout, and the values may be averaged or otherwise manipulated. Interval scaling is always continuous; between any two successive values there may be an infinity of fractions. Examples are degrees Fahrenheit or Celsius, elevation (expressed in feet or meters), and ocean depth (measured in fathoms).

Ratio scaling exists when values of the interval type are based on absolute zero, allowing the determination of true ratios between measurements. The Kelvin system of measuring temperature is an example of ratio scaling that is continuous. Ratio scaling may also be discontinuous (measurable in integers only), the prime example being counts of population.

The difference between interval and ratio scaling can be confusing. For instance, the Celsius and Fahrenheit systems for measuring temperature are common examples of interval scaling. If the subject is relative warmth, 8°C

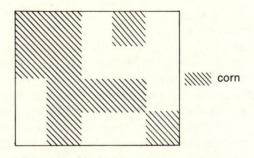

Figure 7–2A Experimental Farm: Presence of Corn

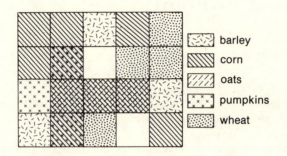

Figure 7–2B Experimental Farm: Presence of Five Crops

is not twice as warm as 4°C, and the scaling is clearly of interval type. If the subject is degrees Celsius above freezing, however, 8°C is twice as many degrees above freezing as 4°C, and the specific reference to freezing (which occurs at 0°C) converts the scaling from interval to ratio type. The passage of time has no true zero starting point, and hence time is represented by interval rather than ratio scaling. However, if the subject is years in our calendar system, then there is a true zero and the scaling is of ratio type.

Cole and King (1968, p. 54), illustrating interval scaling, speak of measurement from a variable zero level as follows:

> A good geographical example is given by altitude above present mean sea-level, which is itself a continually fluctuating zero. It cannot be said that a hill 200 ft. high is twice as high as one 100 ft. high, because if sea-level were to change, this relationship of ratio would no longer hold.

Sea level does, of course, change with varying tide and wind. But any measurement defined as taken from actual sea level at a particular time and place starts from a true zero point, and in consequence would qualify for ratio scaling. Another measurement of sea level taken at some other time or place would also fit the category of ratio scaling—though the two measurements would be equal only by chance. For this reason, when sea level is used as a base in measuring elevation and depth on maps, it is always a mean carefully computed over a long period of time and thus is hardly an arbitrary zero point.

The terms *nominal, ordinal, interval,* and *ratio* are used to indicate what types of statistical tests might be appropriate under varying circumstances, but the distinctions are not significant in thematic map design. Once continuous variables are classed, they receive the same graphic treatment as ordinal variables, which they have in effect become. And, as discussed earlier, the difference between nominal and ordinal values may be just in point of view. Because the terms are frequently encountered in technical literature, however, it is good to know their meanings—if only to avoid being misled about their significance to the cartographic process. The distinction between addable and nonaddable values is far more important to the map designer.

INTRALOCATIONAL VARIABILITY

As we have noted above, the value indicated for each location applies to the whole location. But there is no way of knowing from the values themselves how they may be disposed within locations. For example, if the subject is population by counties in Wyoming, it is impossible to tell whether the population in any county is evenly distributed or concentrated in particular areas, and if so, where. Intralocational variability is thus another form of map generalization.

Depending upon the subject of the map and the specific circumstances, intralocational distributions can vary tremendously. They can be uniform, as in mapping full-pool elevations by reservoirs in the Tennessee Valley. In that case, the map subject reveals the nature of the distribution. Distributions can also be erratic, as would be expected in mapping New York state property taxes. Although little variability might be found when mapping Kansas wheat production, the nature of the particular distribution would be unknown.

Because of this variability, the size of the locations chosen may be very important. This is the reason locations are usually chosen first. In addition, we suggest collecting information about both locations and values by smaller units than the expected final size. Values that are averages over a location, for example, cannot be disaggregated when more detail than was originally contemplated becomes desirable. (See previous section on Addable and Nonaddable Values.)

Increasing the degree of positional detail might be problematic if a great many locations result or if location sizes become very small. In theory, the only limits are the ability to enlarge the map to accommodate the display and the availability of the necessary information. Symbolism will be affected, however, and the appearance of the map will change when locations and their associated values are split, merged, or some combination of both.

Special Knowledge

As a general rule, it is desirable to show a consistent level of detail throughout a study space so that sound comparisons can be made over the map as a whole. It is usually not advisable for the mapmaker to interject personal knowledge that does not apply over the entire study space. Educated guesses about distributions within a location are as risky as those about distributions within the entire space.

When mapping urban social data, for example, a mapmaker may know that a large park exists within a certain location. Using that knowledge would be inappropriate unless all the open spaces within every location were known. However, if one location within a study space were much larger than the others, using special knowledge to achieve a more uniform level of detail might be justified. (Another approach might be to divide the location into smaller zones.) When special knowledge is used for any part of a study space, adequate explanation must be provided somewhere on the map. (See also Chapter 11, the section on Dots.)

INTERPOLATION

Interpolation is the process of estimating quantities lying between specified values. More formally, it is the method or operation of approximating intermediate terms from given terms in a series of numbers or observations, in conformity with the law of the series.

In a study space consisting of a line, if two adjacent locations one mile apart had values of 8 and 12, interpolation would suggest the value 10 for a point halfway between them, or 9 for a point one-quarter of the way from the smaller toward the larger. (The probability of this being the case would be something like 70,000 to 1.) This example describes linear interpolation, taking into account two points (the end points of a line) only. Areal interpolation takes into account the relation of a point to all surrounding points that have assigned values. Those surrounding points are weighted according to proximity, and the value of the new point is computed. Contour lines connecting points of selected constant value to symbolize terrain changes are a familiar form of interpolated display.

Interpolation includes extrapolation, the less reliable process of estimating quantities lying beyond the specified values. As elaboration on a familiar pattern (whether within or beyond boundaries) does not extend knowledge, interpolation can never be as accurate as acquiring more data.

Interpolation may be an estimate of what is most likely at any position within the study space, knowing only the given values. In mapping rainfall in inches, if the value for a location (the rain gauge) were 10 and the values for surrounding locations were all much greater, it is likely that the values within the space between the rain gauge and the surrounding locations would be higher than ten.

Interpolation may also be *diagrammatic*. In mapping urban population for regions consisting of isolated towns and cities, it might be quite appropriate to employ interpolation over the entire study space (as assigned to an area). The interpolated values could have no literal meaning in the areas lying between the incorporated cities. Yet their display might be valuable in showing the trends of urban population over the study space as a whole.

Interpolation with Addable Values

Although it is usually inappropriate to interpolate with addable values, the procedure can sometimes be employed when locations are differentiated. For example, consider an interpolated map titled "Number of Houses by

Counties." If a portion of that map designates the presence of 800–1000 houses, the user would know that despite the position of the contour lines, each *entire* county with a *base center* within that particular portion had between 800 and 1000 houses (Figure 7–3). The value range applicable to the *entire* location can be read from the contour lines on either side of the location base center. Since the center of area B is about 3/4 of the distance between the 800 and 1000 contours, the value applicable to its area may be judged to be about 950. Similarly, the value of area C may be judged to be just less than 1000. The interpolation also suggests the spatial trends of the data across the whole map, which is often the principal reason for using this procedure.

When absolute values are assigned to base centers and symbolism is interpolated among them, the base centerpoints must be shown.* The interpolation symbolism has quantitative significance at these points only, since addable values can be meaningful only in relation to their assigned locations. For example, in mapping congressional representation by states, symbolized by base centers, values interpolated to other positions are only meaningful as a graphic device to reveal national trends. The map user must grasp this

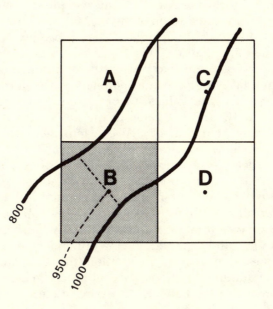

Figure 7–3   Contour Lines with Addable Values, Base Centers Shown

---

*Showing data points when interpolation is used always helps the user judge the reliability of the interpolation. The user should be aware of the number of data points and their spatial disposition, and his trust in the map will be strongly influenced by such information. In exceptional cases where the data point distribution is reasonably uniform and points are very numerous, a written statement might be substituted for display of the data points.

limitation or be confused. Thus the title must be carefully worded: "Congressmen by States, as Interpolated Among State Centers."

Other Considerations

There are many mapping circumstances and subjects for which interpolation is unsuitable. Interpolation is always inappropriate with random values, such as in a map of state lottery winnings per 1000 people by counties. But other examples are less clear-cut.

The scale of the map and the nature of the locations is often significant. When mapping rural population density per square mile over a certain region, interpolated values would obviously not apply to intervening urban areas. But as a graphic device, uniform interpolation might be preferable to interrupting the symbolism at towns and cities.

On the other hand, if one is mapping urban population per acre by blocks, interpolation would certainly be extended across the streets, but could legitimately be interrupted at larger intervening areas such as parks or cemeteries. Locations that are very different in size may also require special treatment. A mapmaker working on subterranean data related to oil exploration would probably not extend interpolation across a major geologic fault in the study space. He might instead divide the space into two independent spaces separated by the fault line. Sometimes barriers to interpolation do not have physical existence. Dividing the space might also be a solution when mapping disease statistics by school districts in townships with different health standards.

One test of suitability for interpolation is as follows: Is it likely that the value for some portion of a location, or in an adjacent space between locations, might be different from the value for the location as a whole? If the answer is yes, the values are suitable for interpolation.

It is sometimes claimed that the number of classes must be related to the number of data points involved: when using interpolation with only a few data points, for example, only a few classes should be employed. This is a misconception. Obviously, the larger the number of data points, and the more uniformly they are distributed, the more accurate the interpolated conceptual surface. However, one can elect to map with whatever degree of precision will be most useful to the map user.

Because of the high cost and degree of skill required in preparation, contour-type symbolism had traditionally been widely employed with physical surfaces only, except for weather maps, and detail was usually minimal— i.e., few levels were used. The advent of computer mapping changed the situation, and the number of levels to be employed is now exclusively a matter of judgment on the part of the mapmaker, dependent on the symbolism available and the degree of detail that will best display the data. A computer-generated map might present as few as two or three levels derived from hundreds of data points, or as many as ten levels from only a few data points.

For a description of a method of hand contouring by linear interpolation, see Appendices 5 and 6.

Questions

1. When is it not essential to show base centers on plane interpolated maps, but likely to be helpful, and why?
2. Under what rather rare circumstances might it be desirable not to show base centers?
3. When is it unsound to interpolate?
4. What are examples of subjects unsuited to interpolation?
5. For routine work within a city planning office would you want to round values generally, or not?

Answers

1. When location base areas vary in size. The centers can serve as useful guides to reliability. It is helpful if base centers are fairly equally spaced.
2. If showing them would attract undue attention, or affect the legibility of the symbolism; or
3. When the location base areas don't really vary in size—and the centers are *extremely* close together. Also, you cannot interpolate (areally) sensibly from elongated locations (roads, etc.), i.e., in linear situations.
4. *Point data* as opposed to *data collected at a point* (rainfall, etc.) is unsuited to interpolation since it is not variable over space.
   Examples of point data:
   a. location of fire stations in New York City
   b. active wells in a county in California
   c. elementary schools in Chicago
   Line data is also unsuitable:
   a. location of major U.S. highways
   b. railroads in Illinois
   Some area data is unsuitable:
   a. location of city parks in Minneapolis
   b. university campuses in Cambridge
5. Rounding should be minimized whenever possible. When rounding is done, detail and information are sacrificed. Presumably, the planning office has spent some effort collecting data at a certain level of accuracy. Rounding tends to make the data less accurate, especially if later manipulations of the data are to be performed. In such situations, premature rounding before data manipulations can magnify the inaccuracies from rounding so that the end result is considerably in error (i.e., locations might be falsely included in higher or lower classes and maps would become misleading). Rounding then, should be utilized only in the final stages of map preparation, and then only with extreme caution.

Proximal Maps

Proximal mapping is a method that uses the principle of linear interpolation to produce a conformant map. The data value for a zone is applied to its center, and every point on the map is given the value of the location center to which it is closest, resulting in a conformant map in which the shapes of the zones correspond to the shapes of the data locations only in exceptional cases. The midpoint between any two data points is determined and lines are drawn connecting all the midpoints, forming a polygon. The value of any point between the perimeter constructed and the nearest data point, i.e., within the polygon, is by definition that of the data point.

For a description of how to construct a proximal map, see Appendix 7.

PART III

Creating the Display

CHAPTER 8

Steps in the Design Process

When all necessary information from the real world is ready for use and map scale and approximate size have been decided, the design process may be carried out. Pick a subject of personal interest and read the following as though you were going to begin mapping. If more than one subject will be portrayed, what follows applies to each.

For simplicity, we will assume that the map is to be hand-drawn on flat paper, but the design problems involved will be essentially similar in all cases. Lightly penciled outlines, dots, or lines may be used temporarily to indicate assignment until all major design selections can be made.*

STEP 1: SYMBOLISM

Make a tentative decision as to which of the following three basic *types of value symbolism* is likely to be most appropriate for the values involved, considering the nature of the study space and locations involved, the scale to be used, and any special circumstances:

— Spot-like (Figure 8–1))
   If symbolism of this type is to be used, each location assignment will be to a point.
— Band-like (Figure 8–2)
   If symbolism of this type is to be used, each location assignment will be to a centerline.
— Field-like (Figure 8–3)
   If symbolism of this type is to be used, each location assignment will be to an area.

*Such pencil work will usually be done on transparent tracing paper laid over a suitable source map or base map, if available.

49

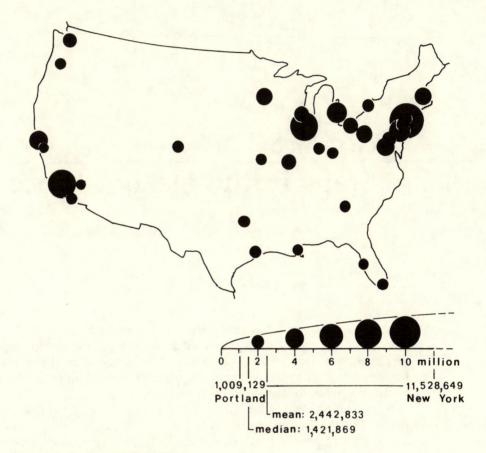

Figure 8–1   Spot Symbolism

With interpolation, the symbolism could be of either band or field type, as when contour lines or oblique views of surfaces are employed.* Locations will be assigned to points. If symbolism based on alternative or subdivided locations is to be used, location assignments will be to areas or points.

To represent applicable values, incorporate one or more of the following graphic *quantitative analogues* within the symbolism selected for use:

—Variation in size or *extent*

—Variation in tone or *darkness*
  (monochromatic or polychromatic)

—Variation in number or *count* (of symbol parts or elements).

Examples of the first category of quantitative analogues are symbolism varying in height, length, width, area or apparent volume—or some combination of these (see Figures 8–1 through 8–3). The most common example of the second category is gray symbolism of varying darkness, usually with some variable texture to aid interpretation (Figure 8–4). Examples of the third category are symbolism consisting of two or more parts or elements,

---

*With oblique (i.e., perspective) views, the symbolism appears to be 3-dimensional.

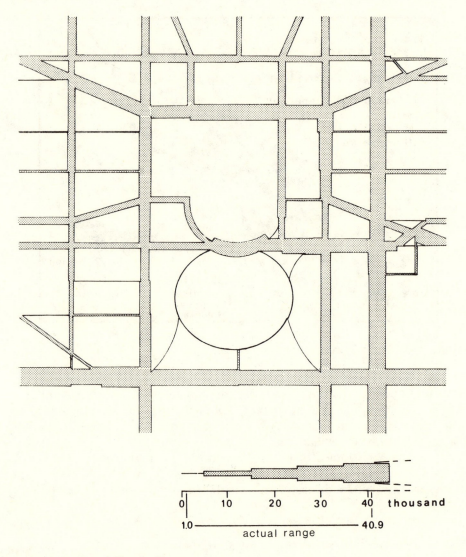

Figure 8–2 Band Symbolism

such as a spot-type symbol composed of a number of dots (Figure 8–5) or a band-type symbol composed of a number of lines (see Figure 1–2.)

All types of symbolism can be used with classed values. Count symbolism among many others, however, cannot generally be used with unclassed values. Some symbolisms can be used for both classed and unclassed values, such as when interpolation is employed with classing defined by contour lines.**

All symbolisms can be used in a map with only a few classes, but only a few symbolisms can be effectively used in a map with numerous classes.

**The individual contour line has a single specific value rating, but the spaces defined by the lines are classed (the lines representing class limits).

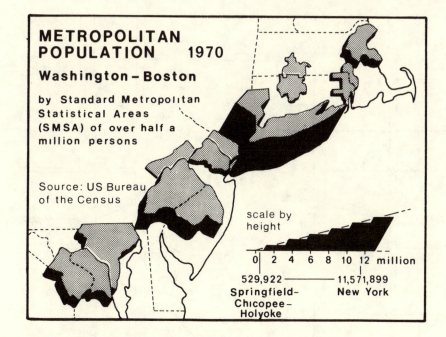

Figure 8–3   Field Symbolism

The number of varying darkness tones that can be successfully differentiated, for example, is quite limited (usually about five or a maximum of ten, depending upon circumstances). A larger number of classes can be successfully used with some symbolisms of count type.

Should the wisdom of the first decision be in doubt later, start over—repeating any of the succeeding steps as required. To assure the best results, it may in any case be desirable to consider more than one possibility.

STEP 2: STUDY SPACE ASSIGNMENT

Assign the study space to an area (2-dimensional) symbolizing the base of its volume. In the process consider the most suitable map projection or transformation to employ.

The study space may not need to be separately symbolized in the final map, since symbolism used for other purposes may serve to define it adequately.

STEP 3: LOCATION ASSIGNMENT

Depending upon the decision made in Step 1 regarding appropriate type of value symbolism, assign each location to a point, a line, or an area.

The locations may not need to be separately symbolized in the final map, since symbolism used for other purposes may serve to define them adequately.

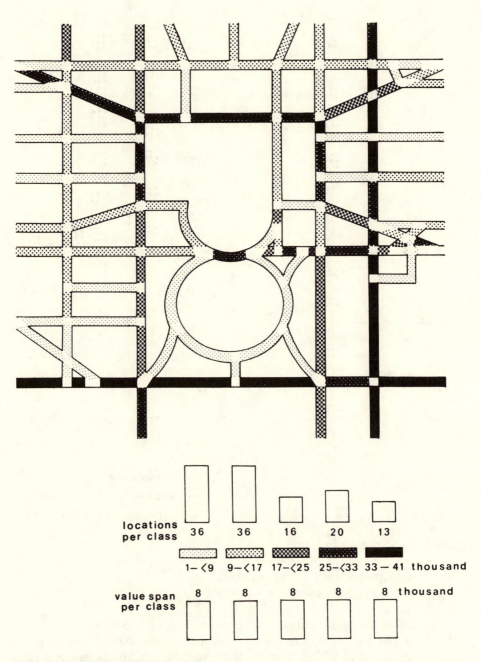

Figure 8–4 **Variation in Darkness**

STEP 4: VALUE CLASSING

For the purpose of symbolizing the values, treat them in one of the three following ways:
   —Unclassed (not grouped into categories),
      applicable to given values of ratio or interval type only

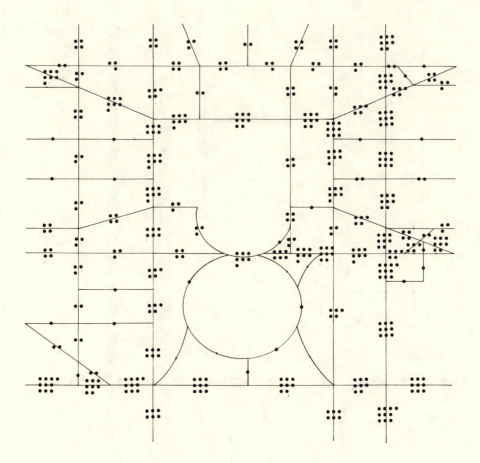

• = 4 thousand

not shown for ⟨2 thousand

(actual range 1,000 – 40,900)

Figure 8–5   Variation in Count

—Classed (grouped into categories),
   applicable to given values of any type
—Mixed,
   applicable to given values of ratio or interval type only.

When classing is used, the class spans must accommodate every given value, with no value falling in more than one class. Gaps between classes should be avoided, since in a sense they constitute additional—though unstated and empty—classes.

With values of ordinal type, which are inherently classed, the given categories may be used by the map designer, or they may be combined to yield a smaller number of larger class spans. (Five classes designated "extra small," "small," "medium," "large," and "extra large" might be combined to yield three classes, "small," "medium," and "large.")

When the values of ratio or interval scaling are classed, the classing may be of equal or unequal spans (in whole or in part). An example of five equal class spans, in terms of a value range of 0–100, would be as follows: 0–<20, 20–<40, 40–<60, 60–<80, and 80–100. (Actually, the last of these classes is larger than the others by an infinitely small amount. However, for all practical purposes, they are equal.) Classes of unequal span may be unequal in a great variety of ways.

When values pursuant to ratio or interval scaling are classed, the *class spans* may be defined in any of these three ways:

—In ratio or interval terms only
—In ordinal terms only
—In both

For example, let us assume that the class spans (in terms of three classes) are 10–<30, 30–<50, and 50–70. Such classing would be in ratio or interval scaling only. If defined in ordinal terms only, the classes might be designated "low," "medium," and "high." Or the two approaches might be used simultaneously—then the classes would be designated "low (10–<30)," "medium (30–<50)," and "high (50–70)." Chapter 10 will treat the subject of classing in detail.

MAPPING A BUSINESS DISTRICT

Problem A:

Assume that you are asked to show in one map how the outer limits of a large and successful outlying retail business district grew and shifted in size, shape, and position over an 80-year period.

Note that concern is only with the extreme limits. If a block within the district had no sales (perhaps it was occupied by a park or public building), assume it to be within the limits.

Within and somewhat beyond the general area that has been occupied by the district, you have information for each city block as to the annual sales volume (in dollars) for each of the years 1890, 1910, 1930, 1950, and 1970.

How would you proceed to get the best map possible?

1. What decision(s) would you have to make to get started—entirely aside from any graphic considerations?
2. Would the resulting map be a single-subject or multi-subject map; and if the latter, how many subjects would be involved?
3. What basic type of quantitative analogue would be best to use?
4. What specific type of symbolism would be best to use?
5. Would your symbolism be classed or not; and if so, what would the classes be?
6. Would using two or more separate maps be better than using the one map called for?
7. If yes, how many maps would you use and what would their titles be?
8. In what respect(s) might two or more maps be better than the best possible single map?
9. In what respect(s) might the best possible single map be better than two or more maps?

Answers

1. The volume of sales per block needed to qualify for inclusion in the district.

   How to deal with the changing value of the dollar.

   How to deal with changing concepts of what constitutes a successful shopping district.
2. Multi-subject, 5 subjects
3. Darkness (with variable tone)
4. Plane conformant, outlined only, with all corners rounded, or perhaps contouring
5. Classed: absent and present
6. No
7. (Name of District) 1890
   (Name of District) 1910
   (Name of District) 1930
   (Name of District) 1950
   (Name of District) 1970
8. Less possibility of confusion where the district may not have grown.
9. The relationships between varying dates would be clearer.

Problem B:

Assume that in addition to showing the extreme limits of the district, you are asked to show the shifting extreme limits of that portion of the district within which the annual sales volume per block was unusually high. How would this affect your answer to each of the above questions?

Answers

1. Same, but also volume of sales needed to qualify for inclusion in high sales area.
2. Same
3. Same
4. Same
5. Classed: 3 classes
6. Yes
7. Two: One for the entire district only, and another for high sales area only.
8. Less confusing
9. Same

Other Considerations

The designer needs to consider two other questions at this point: if the study space is to be directly symbolized, how can this best be done? If the locations need to be differentiated beyond the value symbolism applicable to them, how can this be accomplished?

Compass directions or map scale may be needed, though such information does not involve symbolism in any ordinary sense. It may also be desirable to show latitude and longitude. Are locational clues—such as river, railways, highways, or major institutions—likely to be helpful or confusing? One should not add map cosmetics (locational clues) if the information which is the point of the map is obscured.

FIGURE AND GROUND

Most readers have probably been brought up in the Western world, where teaching from the earliest grades usually involves blackboards; commonly on blackboards the subject is lighter than the ground. We are therefore "ambidextrous" in seeing subject and ground. What are the roles of figure and ground in thematic mapping?

What is intended to be of primary interest, the figure, should immediately appear as such. If the map is concerned with oceanographic data, the ocean area should not appear to be background. A map on white paper of urban housing quality by blocks, using gray-tone conformant symbolism with black streets, has two grounds: the black streets and the white borders around the map. If different gray tones symbolize nonresidential blocks devoted to recreational, commercial, or other purposes, there may be several grounds.

The interplay between figure and ground occurs between two poles, simultaneous contrast and assimilation. A small–grained pattern of equal white and black merges to gray due to assimilation. With a progressively coarser pattern of equal white and black, assimilation changes to simultaneous contrast—the exact point of shift depending upon the viewer's attitude. If one is looking at a large area, the effect of assimilation will persist as coarseness increases. If one focuses on a small area, however, the effect of simultaneous contrast will soon appear as coarseness increases (Figure 8–6A, B, C). This phenomenon is basic to the exploration and development of graphic symbolism.

The best degree of coarseness to use in a conformant map of gray tones permits assimilation when the user's attention is given to the map as a whole or to substantial portions of it, and allows simultaneous contrast when the user's attention is concentrated on a small area. Broad judgment is facilitated by gray tone, while the positive identification of tones of particular interest is aided by the geometry of the pattern assisted by simultaneous contrast. If the grain is too fine (below the resolving power of the eye), only the first objective is served. If the grain is too coarse, only the second objective is served.

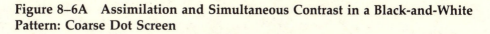

Figure 8–6A   Assimilation and Simultaneous Contrast in a Black-and-White Pattern: Coarse Dot Screen

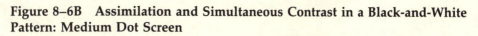

0 – <20 20 –<40 40 –<60 60 –<80 80 –100

Figure 8–6B   Assimilation and Simultaneous Contrast in a Black-and-White Pattern: Medium Dot Screen

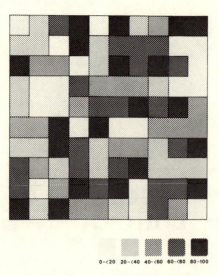

Figure 8–6C   Assimilation and Simultaneous Contrast in a Black-and-White
Pattern: Fine Dot Screen

CHAPTER 9

Varieties of Symbolism: The Foursquare Study

In Chapter 6, we mentioned an imaginary study space, one of great simplicity, named Foursquare. Our first illustrative mapping problem will be set there, allowing us to introduce a variety of symbolisms.

THE STUDY SPACE: LOCATIONS AND VALUES

The Foursquare study space is straight-sided (Figure 9–1). It measures 10 units by 10 units horizontally by 1 unit vertically. The size of the units is immaterial to us. Our study space is divided into 100 locations, each a cube measuring 1 unit by 1 unit by 1 unit. In real life, study spaces will seldom be so symmetrical and regular.

Table 9–1 presents the values for each of the 100 locations. Each value is shown to one decimal place, and it so happens that no two values are the same. In recording values, it is generally desirable to use at least three significant figures if possible. This reduces the likelihood of identical values.

When values are given in a tabulation, each location must be assigned a unique identification for cross-reference. For this purpose integers beginning with 1 are usually desirable, especially if the computer is to be used. The locations in Foursquare are orderly, and we can start the reference numbers at the upper left corner and proceed across and down, the usual sequence employed in reading printed matter. Figure 9–2A shows the base areas of the locations, with the applicable value for each location printed directly on it. Figure 9–2B shows the base centers of the locations with the applicable value for each location printed nearby.

The maps shown in Figures 9–2A and B present in full the values applicable to the various locations. Yet the spatial disposition of the data over the study space as a whole cannot be comprehended. Maps that require

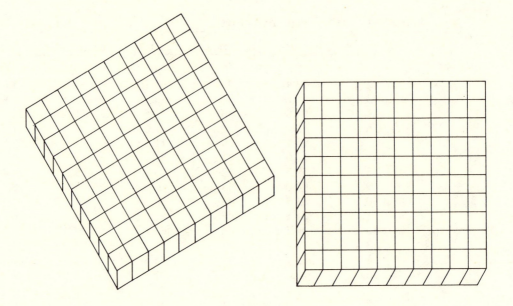

users to focus upon individual locations or small clusters of locations are of little worth for most purposes. If the goal is the communication of precise values, maps are not as efficient as tables.

Let us now employ graphic display to represent the values applicable to each of the 100 locations—through the substitution of quantitative symbolism for the Arabic numerals first employed.

INTRODUCTION TO GRAPHIC SYMBOLISMS

There are many types of graphic symbolism which can be employed by the mapmaker. The challenge is to choose that symbolism which is best suited to meet the unique graphic objectives of each map.

The representation of varying values is usually the most difficult issue, although the problem is minimal in maps involving values of only limited scope (such as those in presence-and-absence maps). The most revealing symbolisms, however, combine good generalization and particularization, but tend to be weak in representing the locations and the study space as a whole.

In facing each new assignment, the wise map designer will make a conscious effort to avoid two pitfalls. The first is the tendency to employ a few types of symbolism under all circumstances; the second is the tendency to shift about among a variety of types for no good reason except novelty. Either tendency will result in inferior productions; the second can also cause confusion and inefficiency when applied to maps in a series.

Table 9–1: The Foursquare Values

Locations	Values	Locations	Values	Locations	Values
1	17.8	35	58.0	68	39.2
2	38.4	36	53.2	69	61.6
3	63.7	37	48.9	70	89.9
4	91.0	38	33.8	71	74.9
5	1.7	39	11.0	72	41.2
6	95.1	40	11.3	73	33.9
7	25.0	41	26.0	74	60.7
8	59.5	42	6.9	75	29.4
9	90.6	43	6.2	76	79.0
10	81.5	44	53.5	77	83.1
11	24.1	45	62.6	78	30.6
12	34.7	46	69.9	79	34.9
13	14.4	47	81.1	80	34.3
14	81.1	48	76.4	81	69.8
15	20.6	49	91.8	82	9.4
16	74.1	50	46.6	83	41.5
17	71.0	51	58.8	84	96.8
18	69.2	52	46.4	85	68.5
19	5.4	53	91.3	86	87.8
20	12.7	54	16.6	87	3.0
21	100.0	55	82.8	88	65.4
22	49.5	56	16.0	89	74.9
23	58.6	57	15.4	90	82.3
24	7.4	58	44.9	91	89.1
25	22.5	59	54.5	92	36.0
26	92.1	60	59.5	93	94.8
27	77.4	61	24.1	94	47.2
28	93.2	62	24.1	95	53.1
29	77.1	63	85.7	96	21.9
30	6.6	64	74.6	97	0.0
31	37.7	65	28.0	98	69.0
32	1.1	66	45.8	99	56.0
33	14.9	67	42.1	100	18.0
34	78.8				

Symbolisms can be categorized by their principal features, positional characteristics, and quantitative analogues. The three basic types of *positional characteristics* are:

1. SPOT type: Symbols that stand alone, completely independent of one another. They are usually relatively compact in shape, and of a single standardized form for any one set of values.
2. BAND type: Symbols that are elongated and usually interconnected. They are of a single standardized form for any one set of values.
3. FIELD type: Symbolism that extends over areas of varying size and shape.

A fourth type consists of symbolism that is actually 3-dimensional in form—a physical solid (sphere or cube). Such volumetric spot symbols are seldom used, however, and we will confine our attention to the three listed above, all capable of being drawn and displayed on paper.

17.8	38.4	63.7	91.0	1.7	95.1	25.0	59.5	90.6	81.5
24.1	34.7	14.4	81.1	20.6	74.1	71.0	69.2	5.4	12.7
100.0	49.5	58.6	7.4	22.5	92.1	77.4	93.2	77.1	6.6
37.7	1.1	14.9	78.8	58.0	53.2	48.9	33.8	11.0	11.3
26.0	6.9	6.2	53.5	62.6	69.9	81.1	76.4	91.8	46.6
58.8	46.6	91.3	16.6	82.8	16.0	15.4	44.9	54.5	59.5
24.1	24.1	85.7	74.6	28.0	45.8	42.1	39.2	61.6	89.9
74.9	41.2	33.9	60.7	29.4	79.0	83.1	30.6	34.9	34.3
69.8	9.4	41.5	96.8	68.5	87.8	3.0	65.4	74.9	82.3
89.1	36.0	94.8	47.2	53.1	21.9	0.0	69.0	56.0	18.0

Figure 9–2A   Base Areas with Values and Location Boundaries

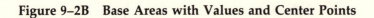

17.8	38.4	63.7	91.0	1.7	95.1	25.0	59.5	90.6	81.5
24.1	34.7	14.4	81.1	20.6	74.1	71.0	69.2	5.4	12.7
100.0	49.5	58.6	7.4	22.5	92.1	77.4	93.2	77.1	6.6
37.7	1.1	14.9	78.8	58.0	53.2	48.9	33.8	11.0	11.3
26.0	6.9	6.2	53.5	62.6	69.9	81.1	76.4	91.8	46.6
58.8	46.6	91.3	16.6	82.8	16.0	15.4	44.9	54.5	59.5
24.1	24.1	85.7	74.6	28.0	45.8	42.1	39.2	61.6	89.9
74.9	41.2	33.9	60.7	29.4	79.0	83.1	30.6	34.9	34.3
69.8	9.4	41.5	96.8	68.5	87.8	3.0	65.4	74.9	82.3
89.1	36.0	94.8	47.2	53.1	21.9	0.0	69.0	56.0	18.0

Figure 9–2B   Base Areas with Values and Center Points

Any of the positional categories may be combined with any of the three types of *quantitative analogues.*

1. EXTENT: ·Symbolisms that vary somehow in size.
2. DARKNESS: Symbolisms that vary somehow in tone, or darkness.
3. COUNT: Symbolisms that vary somehow in the number of countable elements of which they are composed, and which also vary in size.

Usually only one type of quantitative analogue is present. Under some circumstances a symbolism may straddle two or more categories, but in such cases one category tends to be predominant.

The various types of positional and quantitative variables described can be combined in nine different ways, as shown below:

	EXTENT	DARKNESS	COUNT
SPOT	1 SE	2 SD	3 SC
BAND	4 BE	5 BD	6 BC
FIELD	7 FE	8 FD	9 FC

In the following discussion, the numbers identifying symbolisms indicate their position on the symbolism chart, Figure 9–3, and their type.

SEVEN BASIC SYMBOLISMS

Spot-extent: Circles (1-SE3, Given Values, Unclassed)

Let us start with a symbolism that has long been favored for thematic mapping. To represent the value applicable to each location, a circle is employed, usually proportional in area to the value to be represented. This is accomplished by making the radius proportional to the square root of the value.

Circles are best placed on the base centers as shown in Figure 9–4. The range of circle sizes used and the tone employed depends to some extent upon the judgment of the designer. (Note that for the location with a value of zero, the small dot designating its base center has been preserved to indicate the existence of the location.)

This map appears to be reasonably successful, and is certainly far superior to those using Arabic numerals. From the circle sizes we can see at a glance that the values, irrespective of position, tend to be distributed more or less equally throughout the range (0–100). It is even more apparent that the values are disposed in a rather random manner, with numerous discrete high and low value regions. Both distributions are unusual.

If Foursquare represented a district of personal concern, such as a ten-block city neighborhood centered on your place of employment, and the values represented the excess of average housing rentals in dollars beyond your budget, you would certainly find it highly revealing. You would know immediately where to start searching for a new place to live.

In examining each of the alternative maps, start by testing the degree of success achieved in communicating the overall spatial pattern of values before you proceed to look at greater detail. To illustrate, if our circle map represented a proposed office building site and the values represented depth to bedrock in feet (from test borings made at the base centers), a structural engineer would be able to see at a glance that no serious foundation problems existed—after which he could read the information needed to start planning the excavation and foundation work. Similarly, if the promoters of a new ski resort wanted to map an uncharted region and the values showed height of terrain in hundreds of feet, the potential of the region for their purposes would be immediately apparent. They could then identify particular regions with great differences in elevation and determine those likely to combine good accessibility with the greatest variety of runs for a minimum investment in ski lifts.

Circle symbolism has enabled us to gain much information that was previously hidden, but we have also lost information. Even if we attempt to judge each circle separately by checking its size against the value key, we cannot read specific values. It is hardest to judge specific values with the larger circles, where change in area is far less apparent than in the smaller circles. The curve along the top of the value key clearly illustrates this. If the value range were 92–100, it would be all but impossible to make meaningful distinctions (since the circle sizes would be very similar), and the map would be useless. *For the successful use of circle symbolism, the largest value should normally be several times the smallest value.* (As we shall see, this limitation applies to all symbolism of the EXTENT type.)

Note that variability in darkness over the study space is highly revealing and greatly aids overall readability. In fact, the user may be influenced more by relative darkness than by relative circle area. Variable darkness can be misleading, however, if the locations vary significantly in size and the values are nonaddable. In this example the locations are all the same size, and there is no such difficulty; but this problem should always be considered when SPOT symbolism is used.

If the circles are merely outlined, the effect of variable darkness is greatly diminished. This is advantageous when the locations are numerous and vary greatly in size, a situation likely to require the use of relatively large and closely spaced circles. At the other extreme, solid black circles are frequently used for nonaddable values but often look heavy-handed against a white background.

The location base outlines could have been shown but were omitted for simplicity. With all SPOT symbolisms, the number of locations and their general positions can usually be determined from the symbols themselves. If locations vary significantly in size or are of extended shape it is sometimes desirable to show base outlines.

MAP SYMBOLISM

The maps shown on this chart are all single-subject maps based on classed data. Those symbolisms designated by the letter "u" may also be used with unclassed data.

For multi-subject maps, when their use may be warranted, corresponding symbolism may be employed in a variety of combinations.

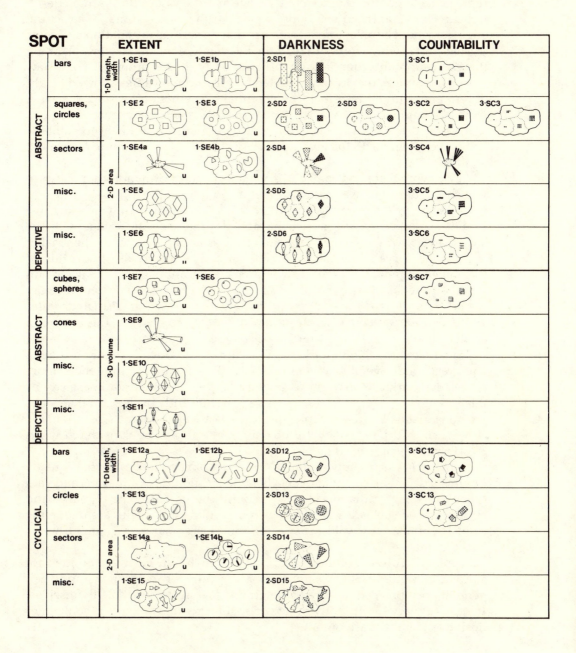

Figure 9–3   Map Symbolism Chart

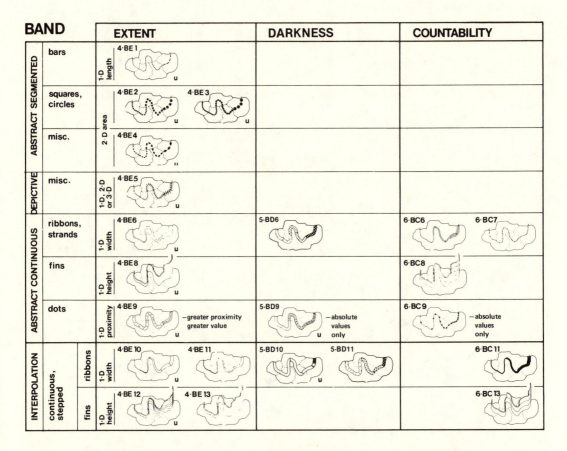

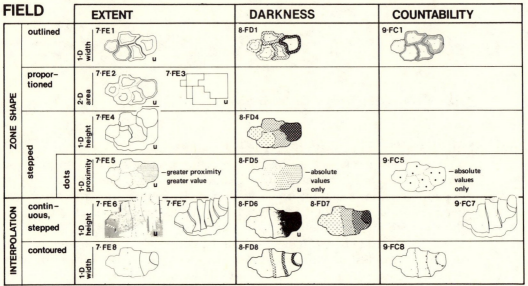

Figure 9–3 **Map Symbolism Chart** (*continued*)

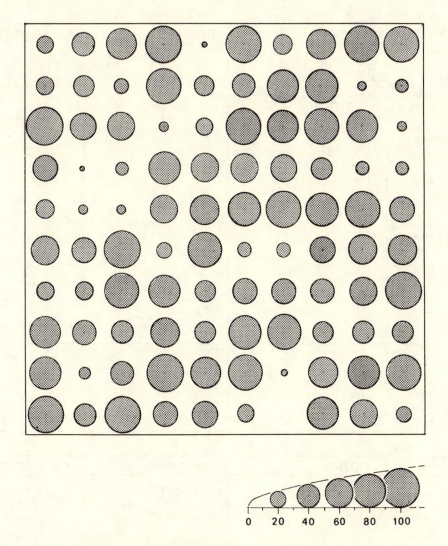

Figure 9–4   Spot-Extent: Circles (1-SE3)

Spot-extent: Bars (1-SE1a, Given Values, Unclassed)

Bars are best placed so that the bottom center of each bar falls on the base center of a location. The range of bar heights, width, and darkness employed depends on the judgment of the designer. In Figure 9–5A the bars have been drawn at an oblique angle so that they will not cross each other, and to give a 3-dimensional effect.

As the height of bars can be judged more accurately and consistently than the area of circles, bars offer some advantages over circles. However, as is true of circles, the relative size of the smaller symbols can be judged with greater accuracy than that of the larger—even though increase in height is directly proportional to increase in quantity represented. As with circles, *the largest value in the range should be several times greater than the smallest.*

The bars here have been drawn solid black rather than gray for better

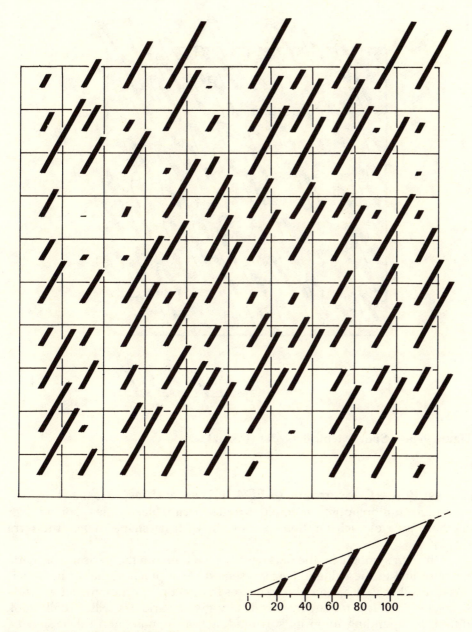

Figure 9–5A   Spot-Extent: Bars (1-SE1a)

legibility. Variable darkness again proves revealing, but less reliably so than with circles, and it can be misleading with nonaddable values when locations vary significantly in size.

In Figure 9–5B we see an alternative form of bar in a map drawn by computer (gray tone added by hand). Here, 3-dimensionality adds visual impact. One disadvantage of this symbolism is that the short horizontal lines representing zero give an impression of there being a value, however slight.

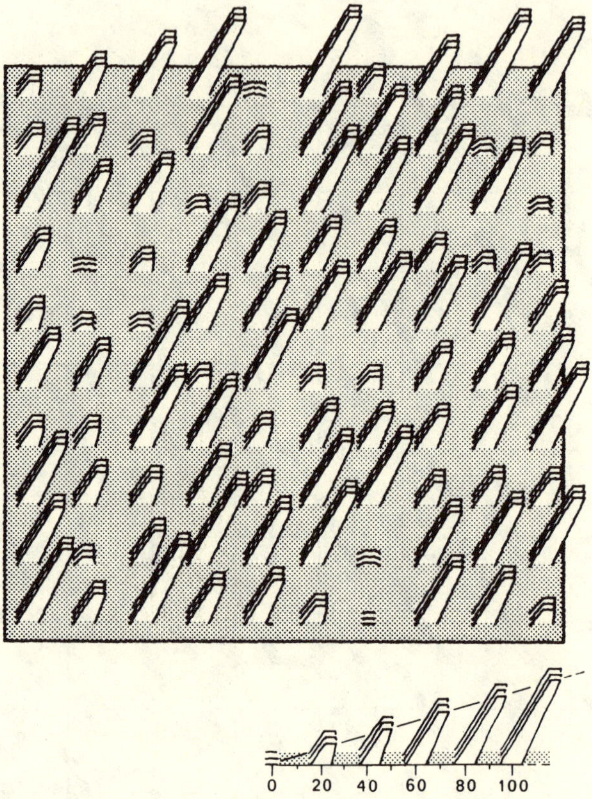

Figure 9–5B   Spot-Extent: Bars, 3D (1-SE1a)

Field-darkness: Plane Conformant (8-FD4, Given Values, Classed)

Another common approach employs tones of variable darkness spread over the base area of each location to conform with its shape (hence the term *conformant*).

With such symbolism, the darkness of each tone can theoretically be made proportional to the value to be represented. As a practical matter however, it is difficult to vary apparent darknesses in direct proportion to value. Aside from the time and money involved, any given darkness will actually look different depending upon its surroundings, a phenomenon called *simultaneous contrast*, and it is difficult for the map user to compensate in advance for this effect.

Therefore, when DARKNESS is used, it is customary to class the data. When classing is employed, the given values are abandoned and the class spans substituted. Each class can be represented by a relatively distinct tone to overcome the adverse effects of simultaneous contrast. The production of maps by hand can be simplified through the employment of preprinted tones composed of black or white dots of varying size at varying spacings, as in Figure 9–6A.

When tones are used with classing they are normally selected for maxi-

0–<20 20–<40 40–<60 60–<80 80–100

Figure 9–6A   Field-Darkness: Plane Conformant (8-FD4)

mum visual differentiability, irrespective of the value spans represented by the classes. This is a powerful argument in favor of classing. If the range of values were 92–100, the same tones ranging from light to dark gray could be used. If tone darknesses were made proportional to the values, however, the tones would be nearly identical and differentiation would be nearly impossible.

Classing represents an extreme form of generalization—with all the possible advantages and disadvantages that implies. With up to five classes the map user can easily understand the values, but it is hard to employ more than five or six classes with gray tones. The designer can minimize the effect of simultaneous contrast by using conspicuous geometric grain within the darkness tones. Compare the spacing used in Figure 9–6A with that of 110 dots per linear inch used in the circle map, Figure 9–4.

When mapping with dot tones, the ideal coarseness is that which gives a satisfactory impression of variable darkness when the map is viewed as a whole, yet allows easy differentiation on the basis of shape among the geometries employed when looking at individual tones in small areas. This balance is usually easy to achieve. The map may then be read generally in terms of darkness, and each class can be clearly identified by symbol shape when examined more particularly.

In the conformant map presented so far, the base outlines of the individual locations only show where different tones meet. When tones of varying darkness are used, showing base outlines in full may present problems. Black lines may not show up clearly when the tone is dark (Class 5 in our example); white lines may not show clearly when the tone is light (Class 1 in our example). Sometimes black and white lines are alternated depending upon the background. This tends to produce a rather disturbing effect, however, and is best avoided. The same problem arises with lettering or other special symbolism, and is a reason for avoiding solid white or black as classing tones.

Figure 9–6B shows a conformant map of darkness type produced by computer. The shape of the individual elements producing the particular darkness tones may be seen. On this map, extra-wide white lines (one character space in width) have been provided to differentiate the various location base areas. When base outlines are shown so conspicuously, it is more difficult to judge the map as a whole in terms of class levels. The emphasis seems to fall upon the locations—their number and equality of size and shape—and comprehension of the values is more difficult than in the prior two maps.

Greater generalization might be helpful under some circumstances. When values shift about as markedly as in Foursquare, it might be desirable to use only three classes, as shown in Figure 9–6C. The tones are far easier to differentiate, and even with the closer element spacing used here we can see the distribution of high, low, and intermediate values more clearly.

Figure 9–6D shows the value distribution of each class separately. Although the ability to show spatial interrelationsips among the classes is entirely lost, separate small maps such as these are frequently useful.

Field-extent: Raised Conformant (7-FE4, Given Values, Unclassed)

Values can be symbolized in conformant fashion by varying the height of each base area in proportion to the value (Figure 9–7). The base plane is drawn at the lowest value. As with circles and bars, classing is not required. This type of symbolism is somewhat similar to SPOT symbolism of the bar type. An oblique angle is again used for an advantageous 3-dimensional effect.

Since the values are unclassed and no two are identical, the outline of each location is primarily shown by the symbolism itself. However, some base areas are partially hidden by locations of higher values, and therefore are undefined. Note that zero values present no special problem.

As previously stated, it is possible to judge variable height with greater accuracy than variable area. In contrast to bars, raised conformant symbolism

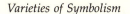

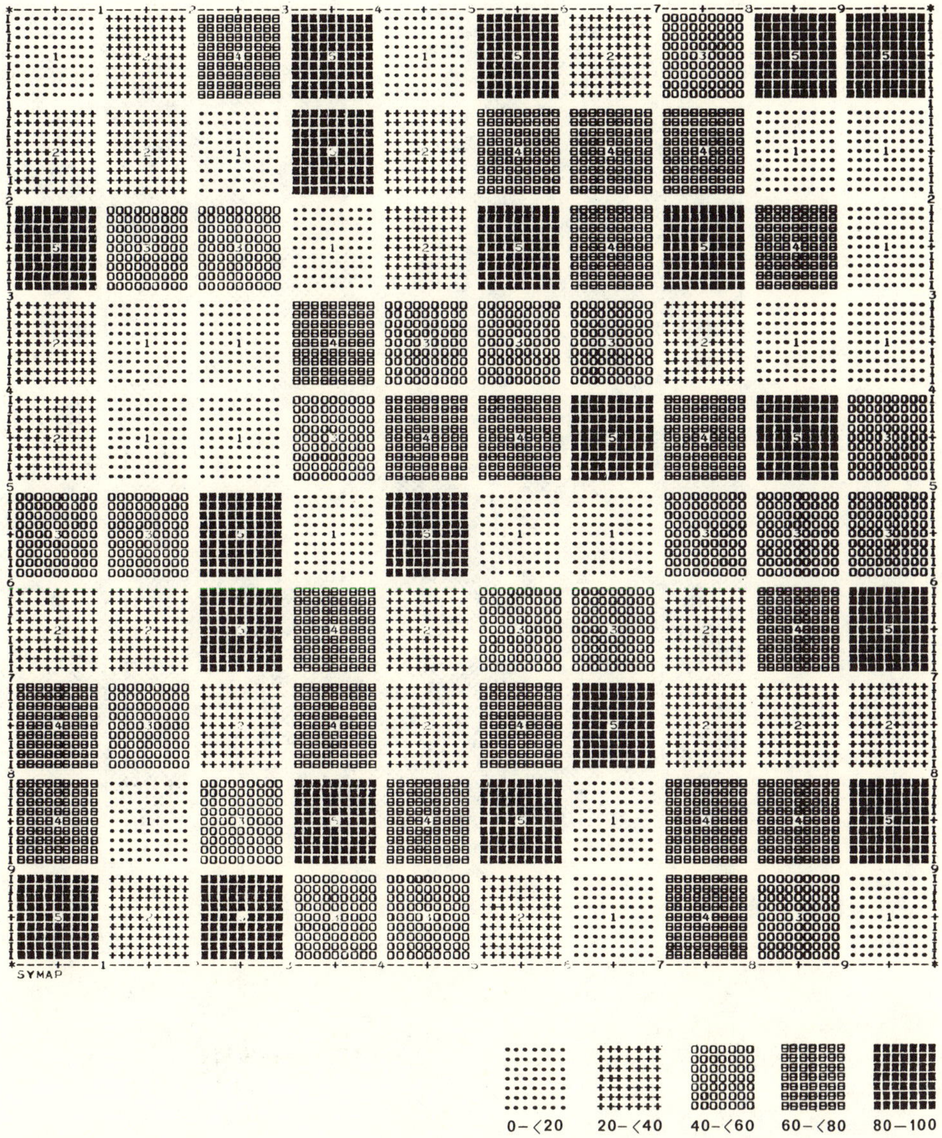

Figure 9–6B   **Field-Darkness: Plane Conformant, Computer-Produced (8-FD4)**

shows the relationships between the values of contiguous or nearby locations more clearly, but makes it harder to judge values individually or to compare values among widely separated locations, since the base plane is hidden except along the bottom and left edges of the study space.

Spot-count: Count (3-SC3, Given Values, Classed)

With this system, quantitative interpretation depends upon the number of elements within each symbol. The elements are disposed so as to create a

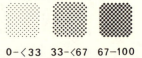

0-‹33   33-‹67   67-100

Figure 9–6C   Field-Darkness: Three Classes (8-FD4)

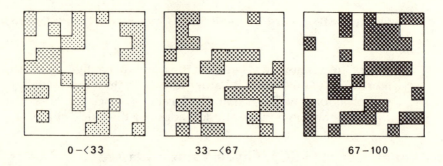

0-‹33   33-‹67   67-100

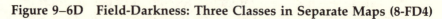

Figure 9–6D   Field-Darkness: Three Classes in Separate Maps (8-FD4)

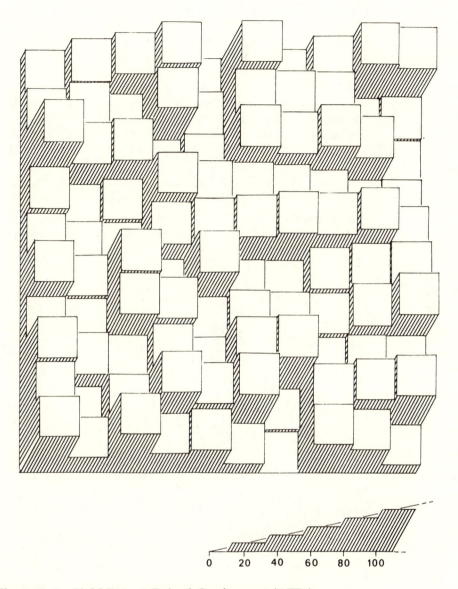

Figure 9–7 Field-Extent: Raised Conformant (7-FE4)

compact symbol shape and facilitate countability (Figure 9–8A). Each element should be a whole number, preferably one that is easy to multiply, and hence classing is again required. A relatively large number of classes can be employed, however, in contrast to the small number that can be used with darkness.

In this example, a single dot element symbolizes 2–<6, two dot elements symbolize 6–<10, and so on through the next to last class of 94–<98. The last class then represents the span of 98–100. There are thus 26 classes, all of the same span except the first and last, which are smaller. Good overall comprehension is possible, yet all values can be read easily within a relatively small tolerance, in this instance, with a maximum error of ± 2. As there are a large number of classes, the span of each class can be quite small, and the generalizing effect of classing is thus greatly reduced.

Readability is enhanced by variation in the overall area of each symbol, although symbol size is not precisely proportional to value. Readability is also influenced to some extent by variable darkness, as with circles. Location outlines may be shown if desired, and symbols should be well separated to avoid confusion. Preprinted dot patterns make production of this symbolism easy.

In Figure 9–8A, the elements making up each SPOT consist of small dots, each representing a quantity of four; however, other shapes and other quantities might have been chosen. While the large number of classes possible is a major feature, SPOT-COUNT symbolism may also be used with a small

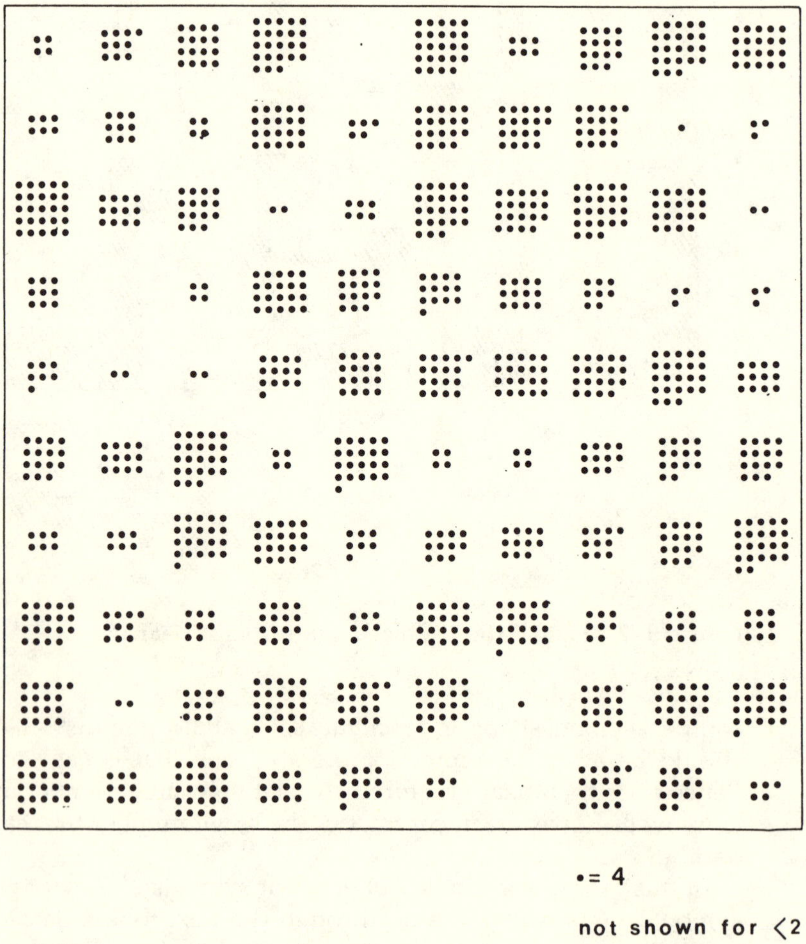

• = 4

not shown for ‹2

Figure 9–8A   Spot-Count (3-SC3)

number of classes. Figure 9–8B shows the use of five classes with square elements arranged into distinctive clustered shapes.* When only a few classes are employed, giving the class span for each separate symbol in the value key is clearer than giving the value per symbol element.

Field-darkness: Dots (8-FD5, Given Values, Classed)

This type of symbolism, closely allied to SPOT-COUNT, spreads dots within each location, instead of arranging them neatly and compactly in a SPOT format. The resulting variable densities yield varying degrees of lightness and darkness, and the map can be easily read, though there is some danger that users may read meaning into the positions of individual dots.

It might be theoretically desirable to arrange the dots for each location uniformly over its base area, since intralocational information to justify other arrangements is lacking, but such a solution is usually impossible. In the square base areas of Foursquare, 9 dots can be arranged regularly, but not 8 or 10. When locations vary in size and shape there is still no reasonable solution.

The next best solution is to arrange the dots as uniformly as possible. Such a solution is illustrated in Figure 9–9A, which has a maximum of 25 dots per location. While the map can be judged broadly through DARKNESS, more detailed interpretation of 26 classes is possible by counting the number of dots appearing within each base outline. This is obviously laborious and this method is therefore characterized as FIELD-DARKNESS rather than FIELD-COUNT.

With this type of symbolism, designers are frequently tempted to omit

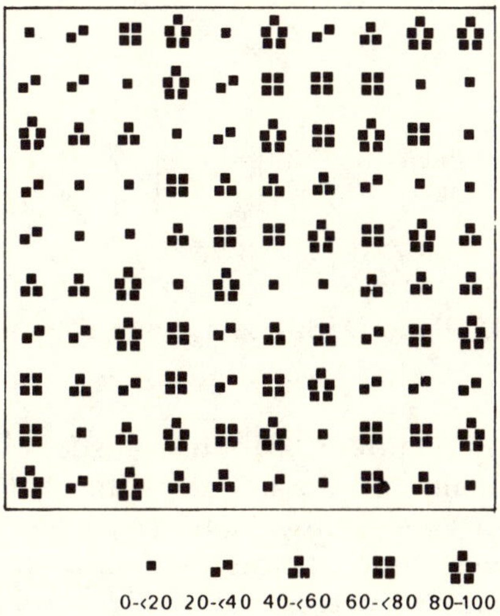

0-‹20 20-‹40 40-‹60 60-‹80 80-100

Figure 9–8B   Spot-Count, Classed (3-SC2)

*Following concepts developed by Roberto Bachi.

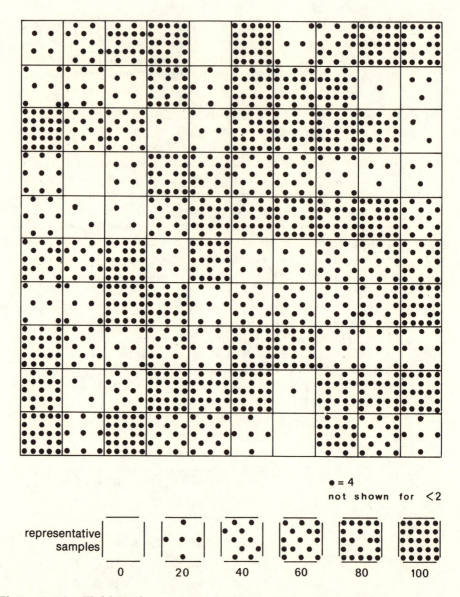

Figure 9–9A Field-Darkness: Dots (8-FD5) with Location Boundaries

base outlines (Figure 9–9B). This is especially likely when the locations are relatively small, and the outlines difficult to show without overburdening the map. In spite of the more attractive appearance thus achieved, two disadvantages ensue: (1) the value for each location can no longer be determined, and (2) the user is more likely to assume that particular dot positions have meaning (as outlying dots in one base area tend to be associated visually with outlying dots in adjacent areas) and the map becomes very impressionistic.

Providing a darkness value key is useful to the reader. If the map also

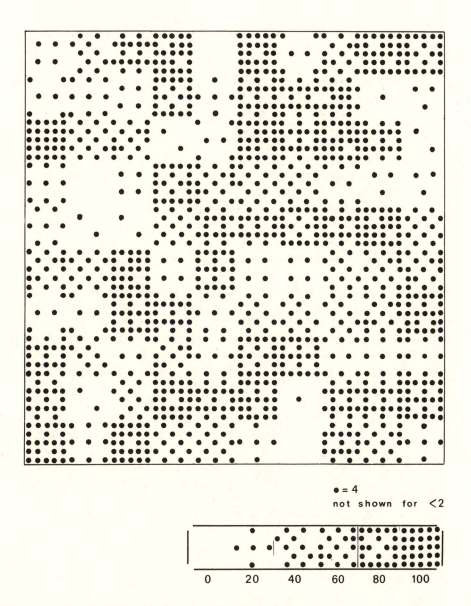

Figure 9–9B   Field-Darkness: Dots (8-FD5) without Location Boundaries

had some specified distance scale, such as "1 inch = 40 miles," it would be possible to represent the values on a "per square mile" basis, with or without base outlines. In this connection it may be noted that while the values used are of addable type, the display also represents nonaddable values in terms of darkness.

Compare the conformant map of Figure 9–6A with the dot map of Figure 9–9A. The concepts are basically similar in that both employ dots and both are read by DARKNESS. The major difference between the two is that in the former, the dots are regularly and closely spaced, while in the latter they are irregularly and less closely spaced. In the former, variable gray is the result of variable dot size, while in the latter it is the result of variable spacing of same-size dots. The former is in five classes, while the latter is in 26 classes. The most notable difference is the fact that the latter is more accurate because there are more than five times as many classes, but less accurate because positions of the individual dots appear to have meaning but have none.

Spot-extent: Sectors (1-SE4b, Given Values, Unclassed)

This is a rather specialized symbolism that can be useful for showing un-classed percentage values when the value range varies substantially. As our Foursquare range is 0–100, we may imagine that our values represent percentages.

Figure 9–10 shows how sectors of variable arc, proceeding clockwise from 12 o'clock, show percentage values. None of the previous symbolisms made this relationship so clear. It is possible with this symbolism to judge values with great accuracy. For example, we know precisely how sectors should appear when representing 25 percent, 33⅓ percent, 50 percent, 66⅔ percent, and 75 percent—and that any arc with a multiple of 6 minutes or 36 degrees will represent a corresponding multiple of 10 percent. No other symbolism shows individual values with this degree of precision—except perhaps COUNT symbolism with numerous classes and base outlines shown. On the negative side, however, the complexity of the shapes and related circles makes overall comprehension somewhat difficult. The sectors have been made solid black rather than gray for maximum readability.

Symbols of sector type are best located on the base centers, and base outlines may be shown if desired. There is no problem with the representation of zero. Readability is influenced to some extent by variable darkness but probably less than in SPOT symbolism of circle, bar, and count types. Overlapping of symbols may be tolerated to some extent.

INTERPOLATED SYMBOLISMS

For each of the seven types of symbolism so far considered, we have used one of two kinds of assignment: the value applicable to each location has been assigned to its base area (as in Figure 9–2A) or to its base center. In the next several examples, the situation will be different. The symbolism will be determined by the values in two or more locations through the use of interpolation between adjacent or among nearby centers.

In the first two examples to follow, the interpolation will be linear—established in a straight line between adjacent base centers. In others to follow, the interpolation will be areal—established over area among a number of neighboring base centers. As we shall see, interpolation symbolisms of areal type can be especially useful to thematic mapping.

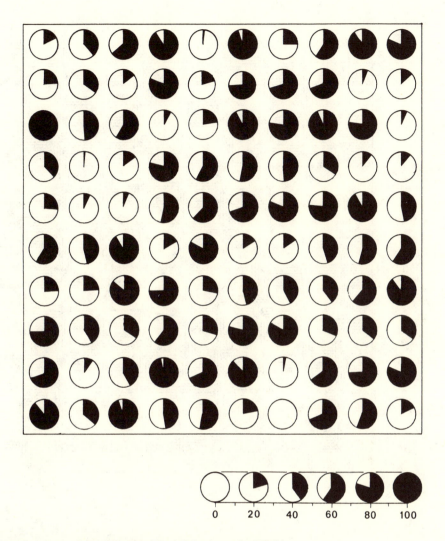

Figure 9–10   Spot-Extent: Variable Arcs (1-SE4b)

Band-extent: Fins (4-BE12, Given and Interpolated Values, Unclassed)

Since the base centers of the Foursquare locations all fall in an orderly pattern of straight lines, it is possible to employ BAND type symbolism, as illustrated in Figure 9–11.

This type of symbolism is similar to bars in that it is judged by height. The base plane is sufficiently visible to permit equally good judgment of height over the entire map. From each base center a line proportional to the given value rises obliquely. Connecting these lines are vertical "fins" that facilitate height comparison between adjacent base centers.

The variable height of the fins conforms to the linearly interpolated values between the base centers. For example, the height of the fin at the halfway

Figure 9–11   Band-Extent: Fins (4-BE12)

point between locations 1 and 2 (with values of 17.8 and 38.4) is halfway between those values (28.1). At any other fractional distance between the base centers, the height of the fin is correspondingly proportional. Thus the fins symbolize interpolated values and their intersections symbolize the 100 given values. Arithmetic computation is unnecessary, since fin height may be established easily by the use of a ruler. Between the outer base centers and the border of the study space the fins have simply been extended at a constant height.

While this type of display is rather complex and has an abundance of detail, it is nonetheless rather revealing. As is typical with SPOT symbolisms, readability is influenced to some extent by variable darkness. In general, where the map is lightest the values are lowest, and where the map is darkest the values are highest. As with bars, however, darkness is not reliable. By running the eye along successive horizontal and vertical fins, added insight can be gained as to the variability of the data. Zero values create no problem, but it would be impractical to show the outlines of locations. With locations of elongated shape, BAND symbolism would normally be positioned along base centerlines, which are seldom straight.

Band-extent: Ribbons (4-BE10, Interpolated Values, Unclassed)

A similar system employing horizontal "ribbons" is shown in Figure 9–12. The analogue for quantity is variable width rather than height, the width being shown in the base plane.

Figure 9–12   Band-Extent: Ribbons (4-BE10)

This symbolism, though simpler and more direct than fins, is less effective in some ways. It is generally more difficult to compare the widths of ribbons than the heights of fins. When values are high, the superimposition or merging of the ribbons at base centers causes problems. Interpolation is performed as in the previous example, except that it is measured equally in two directions rather than just vertically.

Field-darkness: Plane Interpolated (8-FD7, Given and Interpolated Values, Unclassed in Part and Classed by Darkness)

The computer-produced map of Figure 9–13A is based on areally interpolated values derived by the computer from given values assigned to base centers. It illustrates the most widely useful interpolation symbolism. Values should generally be classed with DARKNESS symbolism, and we have here employed the same five equal class spans and darkness elements used for the computer-produced conformant map of Figure 9–6B.

Although all of the values (both given and interpolated) have been classed and symbolized by darkness, the positioning of the symbolism where two tones meet is based on the given values applicable to each location. In other words, the edges of each tone are determined by unclassed data and are unclassed. Thus we have a symbolism which, though classed, provides many of the advantages of using unclassed values.

Within the map border there are 9,801 character locations. The symbolism displayed in each of those character locations is the result of a two-stage process of assignment. In the first stage, the value applicable to each location is assigned to its base center. From these given unclassed values at their various assigned positions, interpolated quantities applicable to the remaining 9,701 character positions are then computed and assigned to positions. Only at this stage are all of the quantities classed and the appropriate symbols printed in each character position. Imagine a change occurring in one of the given values. Even if it were minor or entirely within a class span, the spatial disposition of tone within the map would change.

It is difficult to show location base outlines without an adverse effect upon the quantitative symbolism. However, if the outlines are important for reference purposes, they might be shown on a transparent overlay. The presence of each separate location is evidenced by the white square at each base center. The interpolation process based on these centers varies in reliability with the distance among them, being greatest when the centers are relatively close together. When location base areas vary in size, the positions of their centers can serve as a useful guide to reliability.

Preprinted dot patterns on a hand-drawn map are shown in Figure 9–13B. The tones are spread to correspond with those of the computer map, though the individual dots are not positioned to conform to the character locations of that map. Finer patterns were selected, since with interpolation maps of this kind, adjacent tones are normally only one class step apart. Therefore, problems of simultaneous contrast are minimized and a finer grain tone may be used with confidence.

This type of symbolism offers several major advantages for overall comprehension: (1) it tends to visually organize the data, (2) it combines to a

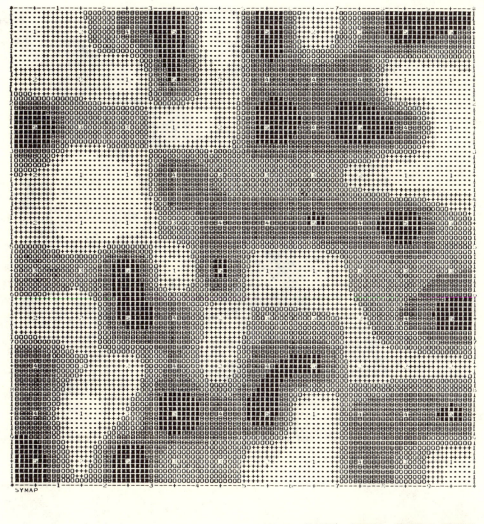

Figure 9–13A   Field-Darkness: Plane Interpolated, Computer-Produced
(8- FD7)

major degree the virtues of both classing and unclassing, and (3) it facilitates
the judging of darknesses.

Field-extent: Raised Interpolated (7-FE6, Given and Interpolated Values, Unclassed)

When the given and interpolated values described in the preceding discussion are symbolized by variable height in a manner comparable to the raised conformant symbolism of Figure 9–7, the result is the computer-produced drawing shown in Figure 9–14A.

0–<20   20–<40   40–<60   60–<80   80–100

Figure 9–13B   Field-Darkness: Plane Interpolated, Hand-Drawn (8-FD7)

Classing is avoided by using height instead of darkness as the basic analogue for quantity. The height of the conceptual surface at every one of the 9,801 positions is in direct proportion to the given values or to the quantities established by interpolation.

With this type of symbolism it is particularly difficult to show the base outlines of the various locations, and an overlay would not solve the problem. The difficulty results from the fact that the base plane is entirely hidden, while the raised surface is undulating. One possible solution is that the whereabouts of each location could be indicated by placing a small asterisk or other symbol on the raised surface directly above the position of each base center.

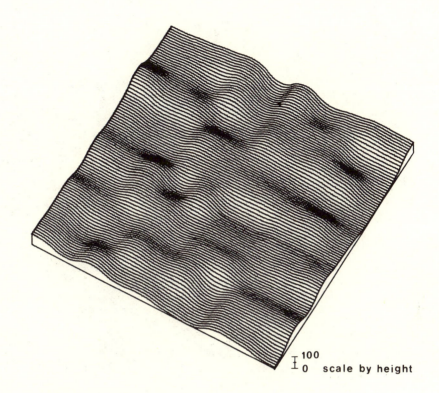

scale by height

Figure 9–14A   Field-Extent: Raised Interpolated (7-FE6)

With such *oblique view* displays, the assumed viewing position and maximum height of the raised surface depend upon the judgment of the designer. In the present example the entire surface has been kept low enough so that no portion of it is hidden from view by any other portion. For comparison Figure 9–14B illustrates three heights of progressively greater magnitude. Up to a point, the greater the height, the greater the ease of interpretation—but at the cost of obscuring some portions of the surface. The designer must decide on the best compromise for any given set of circumstances.

It may be noted that the angle of vision employed here is different from that used in the previous oblique view maps and also that the base outline of the study space has been foreshortened. The previous system is easier to produce by hand, but both types are easily done with the computer. For a given drawing width, however, the area of the display becomes substantially smaller and the study space is no longer *planimetric*, i.e., of true shape. With foreshortening, lines disposed at right angles in the base plane are no longer so represented in the display.

While overall comprehension is usually excellent with this system and each value is shown, it is difficult to relate the values to positions in the base plane. There is no problem in representing zero values, however. In this respect it is similar to the raised conformant symbolism of Figure 9–7.

The lines used to define the surface have all been run in a single direction, from left to right, corresponding with the rows of elements in the preceding

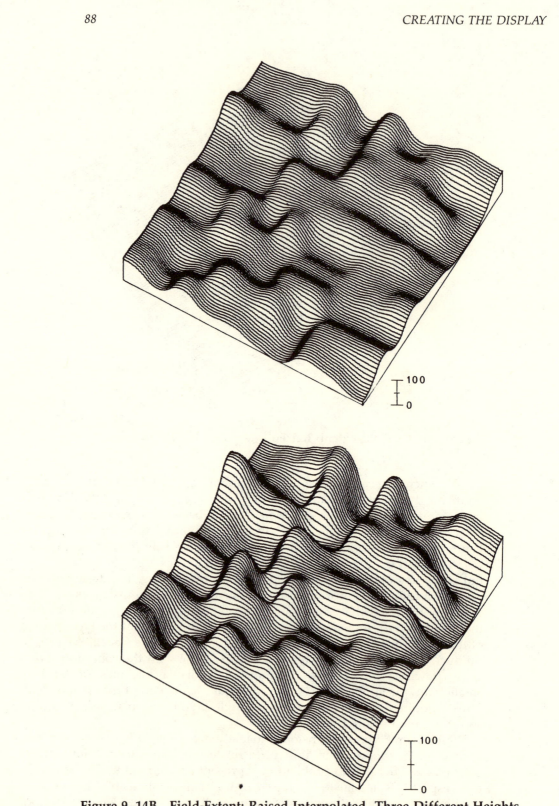

Figure 9–14B   Field-Extent: Raised Interpolated, Three Different Heights (7-FE6)

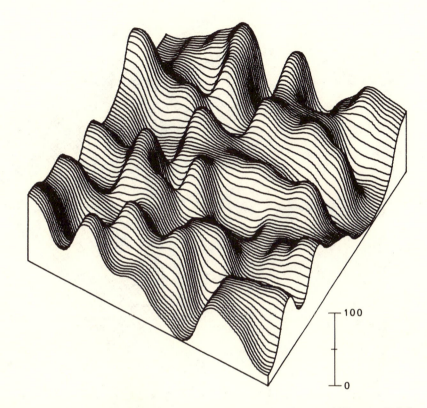

100

0

Figure 9–14B   Field-Extent: Raised Interpolated, Three Different Heights (7-FE6) *(continued)*

computer map (Figure 9–13A). However, they might have been run up and down to correspond with the columns of elements in the computer map, or obliquely along diagonal rows from upper left to lower right or upper right to lower left.

It is also possible to run the defining lines in two directions as shown in Figure 9–14C. When this is done, however, the distance between lines in any one direction must be increased so that the surface will not become overly crowded and dark. A two-way system, showing surface profiles in two directions, offers some advantages, but in general a one-way system of lines closer together will allow for more subtle variations in shading and will provide more information.

CONTOURING

The variable surface displayed in the last several illustrations may, like any undulating terrain, be symbolized by contour lines. Figure 9–15A shows how such contour lines would be disposed, assuming the use of five equal classes. The lines fall precisely wherever the symbolism changed in Figures 9–13A and 9–13B.

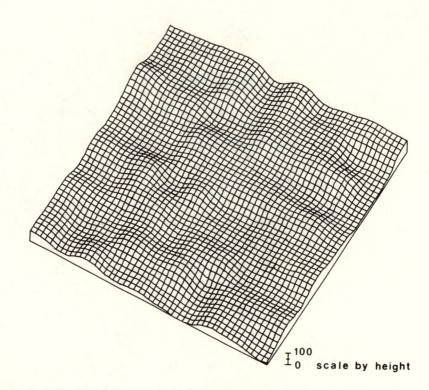

Figure 9–14C   Field Extent: Raised Interpolated, Gridded Surface Shading (7-FE6)

Without any clue to values such a display is meaningless. The contour lines tell the map user where certain constant values lie, but do not tell what the values are. To solve the problem, the particular value of each contour line must be revealed, which is usually accomplished by employing Arabic numerals (Figure 9–15B). Unfortunately, this solution is not graphically readable.

If the number of classes were reduced to three, as in Figure 9–15C, comprehending the map would still be difficult. There is even a problem with only two classes, the fewest possible (Figure 9–15D). The spatial pattern cannot be comprehended without graphic symbolism.

The reader may find the last-mentioned maps too schematic, and feel that more detail might assist comprehension. Figure 9–15E shows the same surface defined by twenty classes. Arabic numerals have been provided for every fourth contour, to avoid overcrowding the map. But providing more information does not solve the basic problem resulting from the use of numbers.

Under most circumstances, providing appropriate darkness tones between the contour lines constitutes the best solution to our problem. This procedure is illustrated for five, three, and two classes, respectively, in Figures 9–15F, 9–15G and 9–15H.

Compare the last three maps to Figures 9–15B, 9–15C, and 9–15D. Figure

= 20, 40, 60, 80,

Figure 9–15A   Contouring with Five Classes (7-FE8)

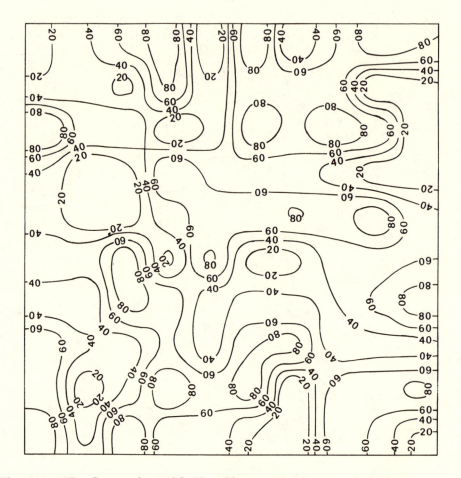

Figure 9–15B Contouring with Five Classes, Numbers Added (7-FE8)

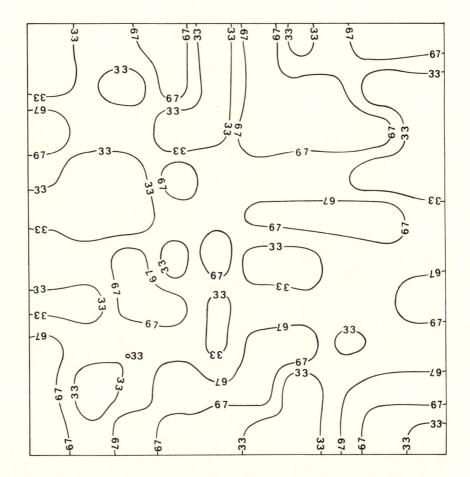

Figure 9–15C   Contouring with Three Classes, Numbers Added (7-FE8)

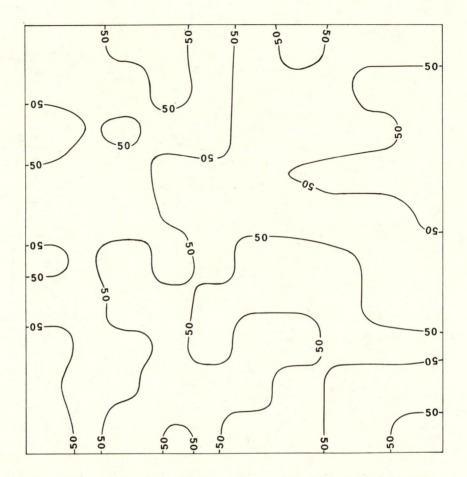

Figure 9–15D   Contouring with Two Classes, Numbers Added (7-FE8)

Figure 9–15E   Contouring with Twenty Classes, Numbers Added (7-FE8)

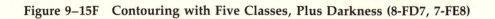

0–<20   20–<40   40–<60   60–<80   80–100

Figure 9–15F   Contouring with Five Classes, Plus Darkness (8-FD7, 7-FE8)

0-<33   33-<67   67-100

Figure 9–15G   Contouring with Three Classes, Plus Darkness (8-FD7, 7-FE8)

0−<50 50−100

Figure 9–15H   Contouring with Two Classes, Plus Darkness (8-FD7, 7-FE8)

9–15F is the same map as Figure 9–13B, but in the latter no contour lines were provided where dissimilar tones met, since at that stage of our discussion the portions of the map falling within each class were emphasized rather than the lines of separation between sequential classes. The combined use of contour lines and tones undoubtedly results in the most easily readable display. Emphasis properly falls upon the tones displayed rather than upon the contour lines, since the former make the map readable.

Field-extent: Raised Interpolated, Contoured (7-FE6, Given and Interpolated Values, Unclassed and Classed by Contours)

Contour lines can also be added to raised interpolated maps of EXTENT type. Figure 9–16A shows such a map, with contouring provided for five classes. By this procedure, we see for the first time all of the given *and* interpolated values, unclassed *and* classed.

Adding contour lines in this case is revealing in a fundamental sense, and is certainly not a matter of detail, as it was with the plane conformant map of DARKNESS type (the preceding example).

In Figures 9–16B and 9–16C we see contouring provided for three and two classes, respectively. Compare the two maps presented here with Figure 9–14A. It would be impractical to attempt to show location outlines here or to employ an overlay.

Field-extent: Raised Interpolated, Contoured (7-FE7, Given and Interpolated Values, Unclassed in Part and Classed by Steps)

A closely allied type of interpolated symbolism creates a terraced effect (Figure 9–17A). The outer edges of the terraces conform to the contour lines for the lowest value in each class, and are hence unclassed.

Figure 9–17B shows a more generalized solution in terms of three classes. Here, it would be impractical to show location outlines or to employ an overlay for that purpose.

THE VALIDITY OF INTERPOLATION

In thinking about interpolated symbolisms, a question naturally arises as to the accuracy, or at least the validity, of the basic interpolation procedure.

contours at 20, 40, 60, 80   100
                                    0   scale by height

Figure 9–16A   Field-Extent: Raised Interpolated, Contoured, Five Classes (7-FE6, 7-FE8)

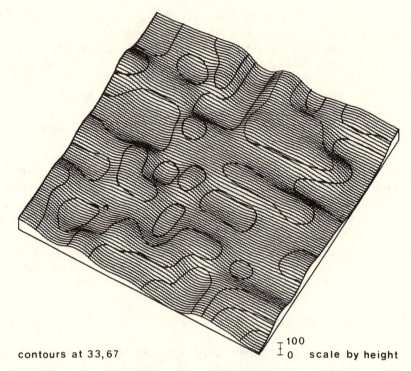

contours at 33,67 $\overline{\underline{\text{I}}}$ 100   scale by height
                                     0

Figure 9–16B   Field-Extent: Raised Interpolated, Contoured, Three Classes (7-FE6, 7-FE8)

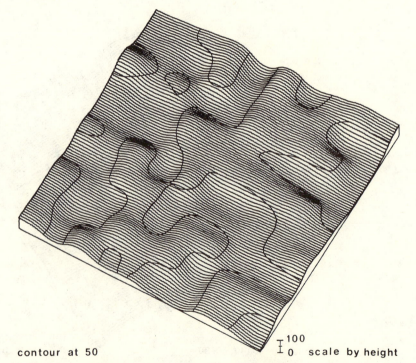

contour at 50 $\overline{\underline{\text{I}}}$ 100   scale by height
                                    0

Figure 9–16C   Field-Extent: Raised Interpolated, Contoured, Two Classes (7-FE6, 7-FE8)

0–<20   20–<40   40–<60   60–<80   80–100

Figure 9–17A   Field-Extent: Raised Interpolated, Stepped, Five Classes (7-FE7)

To cast some light on this matter, let us compare a conformant stepped surface such as Figure 9–7 with an interpolated undulating surface such as Figure 9–14A. The former type of conceptual surface, like the corresponding plane conformant symbolism of Figure 9–6A, is generally accepted without question.

Both of these conceptual surfaces are derived from precisely the same data, and both involve the somewhat questionable process of assignment— to base areas for one and to base centers for the other. The conformant type assumes that the value applicable to each location applies uniformly over the entire base area of each location. The interpolated type assumes that the

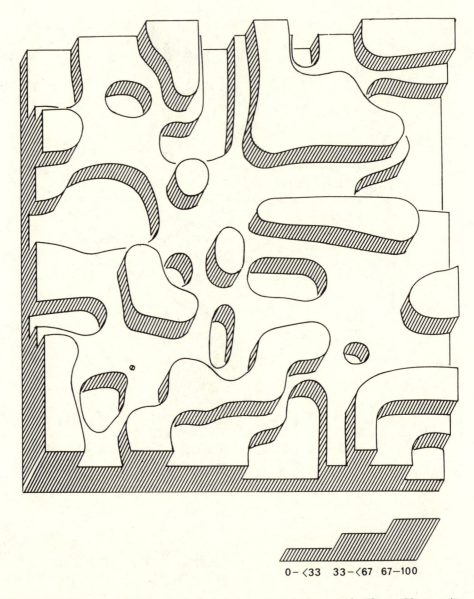

0– ⟨33   33–⟨67 67–100

Figure 9–17B   Field-Extent: Raised Interpolated, Stepped, Three Classes (7-FE7)

value applicable to each location applies to the centerpoint of the base area of each location but not elsewhere within the location, and that the facts applicable to nearby locations are useful clues to that variability. In both cases it may be argued that if the facts are not as represented, it is at least reasonable to so symbolize them. Both assumptions agree on one significant point: in the absence of other information, the best value to assume at the center of a location base area is the value applicable to the location as a whole.

Under neither assumption, except by the most unlikely of coincidences, would the resulting display conform to the true spatial disposition of the data within each location base outline. In general, however, the superior logic of the interpolated approach seems inescapable. By way of example, if a location with a certain value were surrounded by six other locations each with the same value, it would be most reasonable to assume a level surface over the base area of the central location. If the values of the surrounding locations were all much lower, however, it would be more appropriate to assume a convex surface over the base area of the central location.

The difference between the conformant and interpolated approaches may be illustrated by drawing sample profiles through the two types of conceptual surface. If we pass vertical planes through those base centers having the extreme values of zero and 100, the resulting profiles would appear as shown in Figure 9–18. The "true surface" might, of course, be similar to or highly dissimilar from either or both of these profiles, but for most subjects there is a great likelihood that the interpolated profile will prove to be more accurate of the two alternatives.

What has been stated about the two raised surfaces just compared is true regardless of the particular type of symbolism employed. Take, for example, the plane conformant and plane interpolated maps of Figures 9–6A and 9–15F. The disposition of the darkness tones in the latter, especially since it is unclassed, will almost certainly approximate the true facts more closely than the disposition of the darkness tones in the former. In any case, the smoothly flowing curves and sequentially related tones of interpolated maps are usually far more easily comprehended than the angular lines and disjunctive tones of the conformant maps. The conformant map of Figure 9–6A offers a multiplicity of facts generalized by classing but otherwise unrelated, except that adjacent locations falling in the same class share the same darkness tone. In contrast, the interpolated map of Figure 9–15F organizes and relates the variable quantitative information into a more comprehensible, more easily grasped display.

Conformant symbolism frequently draws attention away from the values being represented. Yet under most circumstances, the values in their relative positions are far more important than the particular base outline involved. In those situations where locations must be differentiated, a small symbol or the class number may be placed at each base center. When it is necessary to show base outlines in full, however, interpolation symbolized by darkness or height is not suitable for use. And for some subjects, interpolation is always inappropriate (see Chapter 7).

Our objective is not to recommend the use of interpolation, but rather to counteract a common but incorrect belief that conformant symbolism is inherently accurate while interpolated symbolism is somehow of questionable validity. Both systems are only conventions. The choice between them must always be made in light of the specific circumstances. If the assigning of values to location base centers (as required for the interpolation process) requires justification, remember that spot symbolism, the most time-honored of all thematic mapping symbolisms, likewise employs assignment to base centers. And the assignment of a value to a base area in conformant symbolism automatically includes assignment to its center.

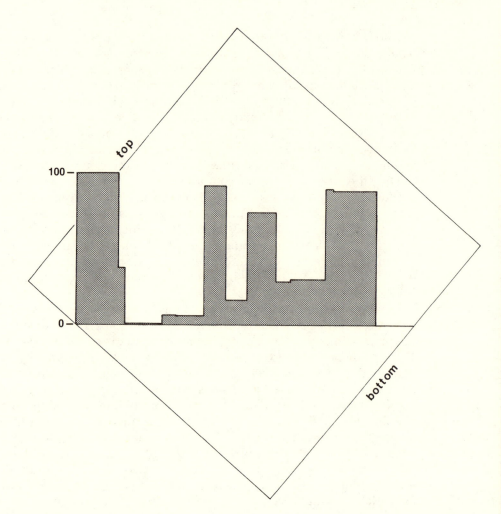

Figure 9–18   Comparison of Conformant and Interpolated Profiles

COLLABORATIVE SYMBOLISMS

Other collaborative combinations than those shown in Figures 9–16A, 9–16B, and 9–16C are possible, and we will illustrate several that allow good general and detailed comprehension.

Figure 9–19A illustrates the combined use of circle and darkness symbolisms to show both unclassed and classed values simultaneously. Circle area symbolizes the given values unclassed, while darkness symbolizes the given values in five classes. Note the two-part value key required.

Figure 9–19B shows the collaborative use of circle symbolism and contour lines for three classes. The greater the width of the line, the greater the value represented. Again circle area symbolizes the given values unclassed, yet

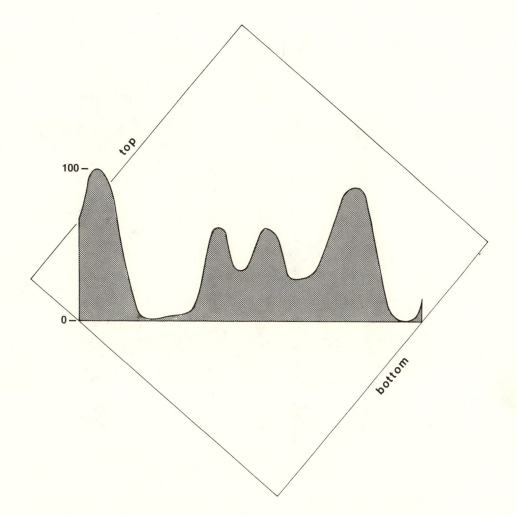

Figure 9–18   Comparison of Conformant and Interpolated Profiles
(continued)

the contour lines aid the map user to sort them out into three broad categories. More than three classes might have been used but the map would
probably have become too detailed.

The circles were made slightly smaller than those used in Figure 9–4, and
the contour lines were shifted slightly from their positions in Figure 9–15C
in order to avoid conflict between the circles and the lines. Compared with
the plane conformant and plane interpolated symbolisms in three classes
(Figures 9–6C and 9–15G), this map provides abundant detail. It is, however,
less easy to read overall than either of those maps or the circle map of Figure
9–4. The two types of symbolisms used here do not cooperate visually with
complete success. A similar approach is illustrated in Figure 9–19C. Here
the background of the intermediate class has been toned in gray, and the
contour lines have been omitted. The result seems somewhat more suc-

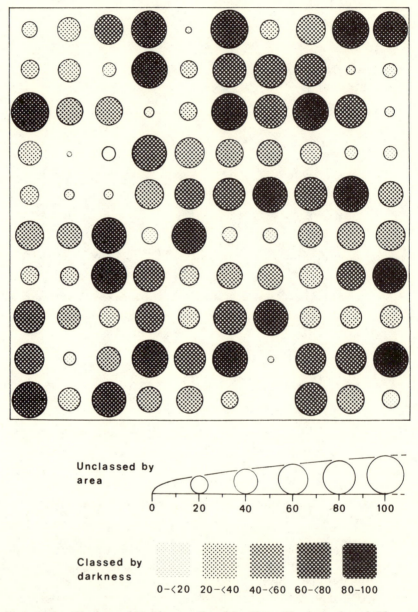

Figure 9–19A   Circle and Darkness Combination, Unclassed and Classed (1-SE3, 2-SD3)

cessful. In a simpler mapping situation with fewer locations or a more orderly change in values, both of the approaches illustrated here might be more successful.

Figure 9–20 illustrates the combined use of bar- and count-type symbolism. Bar height symbolizes the given values unclassed, while the countable bar segments reveal the class within which each value falls. With this sym-

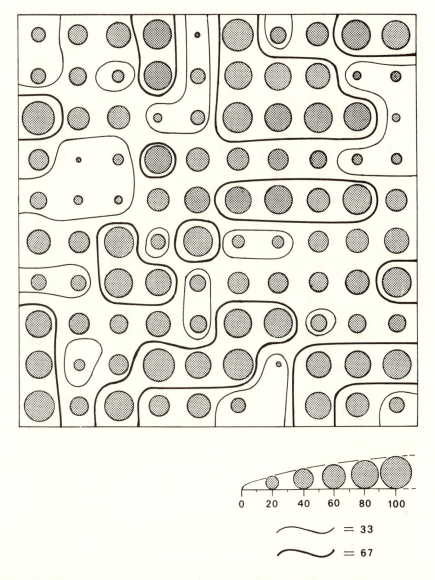

Figure 9–19B   Circle and Contour Combination, Unclassed and Classed (1-SE3, 7-FE8)

bolism it is possible to determine the individual values far more precisely than with ordinary bars or with most other symbolisms. Yet with Foursquare's 100 locations this map is relatively difficult to interpret overall. Again, in a simpler mapping situation this symbolism would probably be more successful.

Figure 9–21 illustrates the collaboration of plane conformant (darkness) and raised conformant symbolisms. Height symbolizes the given values unclassed, while tone symbolizes the given values in five classes. Full detail is provided as in Figure 9–7, but the tone improves overall comprehension.

Figure 9–19C   Circle and Darkness Combination, Unclassed and Classed, No Contour Lines (1-SE3, 8-FD7)

Figure 9–22, though fairly simple in appearance, shows a complex mixture of approaches. It is similar to the raised interpolated, stepped symbolism of Figure 9–17B, except that darkness tones have been added to improve comprehension of height. The map is basically classed by steps and darknesses, while the unclassed horizontal positions of the various step edges are derived by interpolation from the given unclassed values.

Maps such as these, employing collaborative symbolisms, are time-consuming to design and produce by hand, but under certain circumstances they may be well worth the extra effort. In most cases, however, two separate maps with wisely chosen symbolisms will be far more revealing than any single map, no matter how skillfully designed. In terms of Foursquare, the plane interpolated map of Figure 9–15F and the count map of Figure 9–8A might prove highly effective in combination.

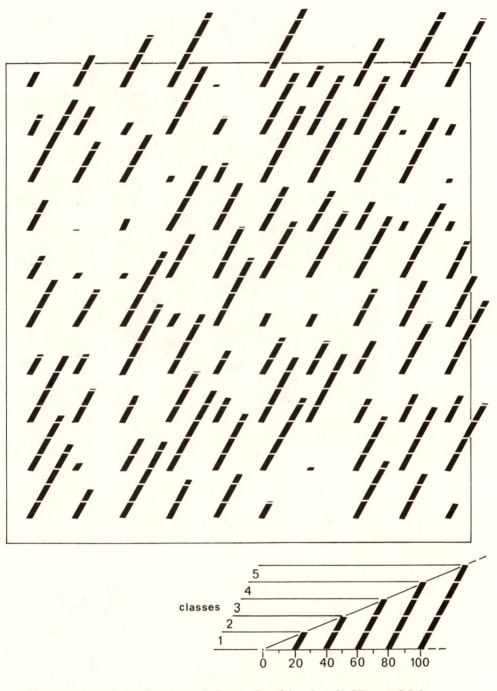

Figure 9–20   Bars: Extent and Count Combination (1-SE1a, 3-SC1)

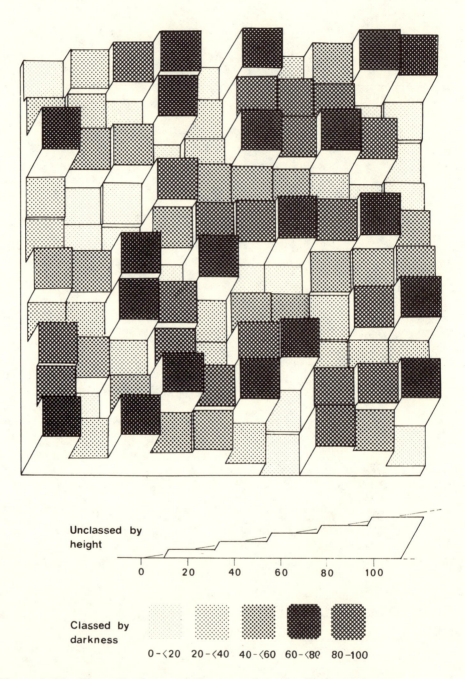

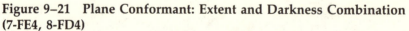

**Figure 9–21   Plane Conformant: Extent and Darkness Combination
(7-FE4, 8-FD4)**

When using collaborative symbolisms, special care should be taken to
ensure that the relationship between the symbolisms is as clear as possible.
This goal may not always be easy to accomplish. An informative title and a
suitably designed value key are important.

It may not always be easy to justify the time, money, or space required

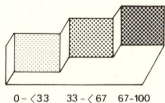

0 – ⟨33   33 – ⟨67   67-100

Figure 9–22   Field-Extent: Raised Interpolated, Stepped, Plus Darkness (7-FE7, 8-FD7)

by multiple maps for the display of any single subject. But when excellent comprehension is important, mapping a given subject in different ways may prove highly effective—as when dealing with a difficult city planning problem, or when trying to decide the best location for an administrative headquarters or a warehousing facility. Under such circumstances, the use of alternative maps may pay dividends which more than compensate for the investment of resources necessary for their production.

CHOOSING THE BEST SYMBOLISM

We have now surveyed a wide variety of solutions to our illustrative mapping problem. Which is best? For the assumed study space, its locations, and its set of data, can we make a reasoned judgment as to the relative merits of the various solutions? Probably not, unless we are prepared to make rather specific assumptions about the circumstances of use.

Each of the solutions illustrated possesses certain virtues and limitations. To choose intelligently among them in any specific instance, we would have to know not only the intended function of the map, but also the type of subject matter being represented, the nature of the values, the shape of the study space, the number of locations and their relative size, shape, and disposition, and so on. A study space with only a few locations, such as New England by states, and a study space with many locations, such as the United States by counties, present very different design problems. Similarly, the choice of mapping method is influenced by considerations such as the desirability of showing given unclassed values (with or without interpolated values), using classing, or perhaps a combination of these. Value curves and classing curves, to be discussed in Chapter 10, can serve as aids in the selection of the best symbolism.

VALUE KEYS

Good value keys are critical to understanding map symbolism. With unclassed symbolism, include the actual values for sample sizes. Design the key so that the viewer can interpolate between samples to closely estimate the value of any symbol on the map. In the preceding Foursquare maps, actual maximum and minimum values (0, 100) were shown (see Figures 9–4 and 9–5A). When the actual data extremes differ from those represented in the key, the true limits should be identified (see Figures 9–23 and 9–24). In the following chapters, we will show the additional information required for classed symbolism keys.

It is advisable to attach a note to most value keys, unless they are *absolutely* clear. Most designers tend to include a note only when a problem is recognized. But even circle symbolism should have a note "judge by area," "judge by height," "judge by darkness," "judge by count," "judge by arc (width)" or whatever applies (see Figure 9–24). With addable values, a note should be inserted in the value key: "Judge by height at data points only."

All value keys should begin at the low end and continue in one consistent direction. The key must be rational, sensible, and memorable, so that it does not have to be referred to constantly by the user.

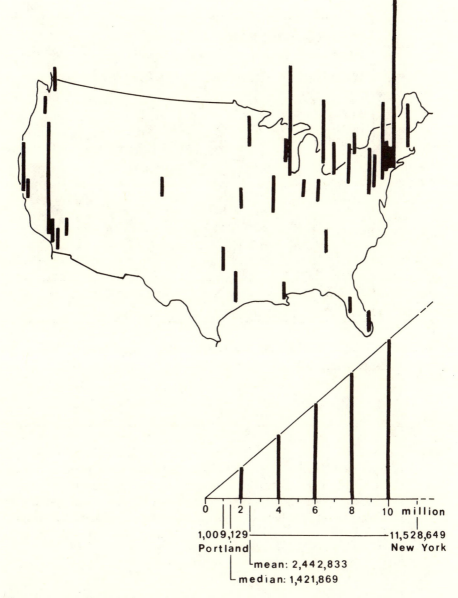

Figure 9–23   Population of Thirty-three Largest SMSAs (Population Over 1 Million Each, 1970)

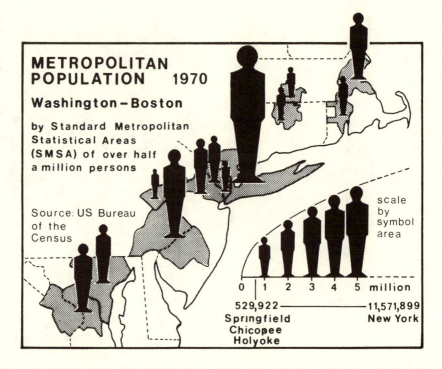

Figure 9–24   Ambiguous Symbolism (Key Must Inform User to Judge by Area, Not Height)

Exercises

For each example listed below, appraise the appropriateness of the following symbolisms. After each, enter any comment you think significant, justifying and if necessary qualifying your choices. State the biggest disadvantage of your first choice. Give your full suggested title.

( ) circles
( ) bars
( ) conformant
( ) conformant, stepped
( ) count
( ) dot
( ) sectors
( ) fins
( ) ribbons
( ) plane interpolated
( ) raised interpolated
( ) contouring
( ) raised interpolated, contoured
( ) raised interpolated, stepped

1. Your study space is the watershed of the Tennessee Valley. Your locations are the numerous reservoirs, large and small, on the various tributaries of the Tennessee River. Your values are the "full pool" elevations in 1950.
2. Your study space is Cambridge; your locations are local streets (i.e., not limited-access traffic ways); your values are related to snow emergencies in two levels: streets judged necessary to keep open, and all other streets.
3. Your study space is Greater Boston; your locations are subway and elevated stations; your values are fares paid in 1980.
4. Your study space is Massachusetts; your locations are municipalities; your values are population per square mile in 1980.
5. Your study space is New England; your locations are states; your values are population in 1930.

CHAPTER **10**

Value Classing

INTRODUCTION TO CLASSING

We have now come to the point where the subject of classing must be dealt with more specifically. The issues posed by the questions of whether or not and how to class are significant and fundamental. They must be carefully explored; for the final map and its message are powerfully influenced by these decisions.

To class data is to arrange it in groups or classes pursuant to some categorizing concept. For example, let us assume that we wish to map the percentage of houses possessing built-in heating systems in southern cities. For each data location (perhaps city block) the actual percentage figures could be employed. If so, the data values and the resulting map would be considered unclassed. Alternatively, the data values might be clustered together into five, ten, or some other number of classes. If so, the resulting map would be considered classed. The classes might all be of the same size or span, or they might be of different spans. The term *class span* will be used to refer to the extent of a class, i.e., the total range of values embraced within its limits, and *zones per class* will refer to the number of zones which might fall within its span.

In classing the data of Figure 9–6A, Foursquare, we used five equal classes: 0–<20, 20–<40, 40–<60, 60–<80, 80–100. It is inadvisable to specify class spans by such designations as 0–19.99, 20–39.99, and 40–59.99. In addition to introducing complex numerals, such a procedure leaves gaps between classes. Gaps of any kind between classes, no matter how small, are best avoided. The class spans should embrace every value from the lowest to the highest *inclusive*. Skipping class spans by designating the value at which the breaks occur is permissible, provided it is clear in which class the values fall. For our example, the class specification would read: "0, 20, 40, 60, 80, 100— intermediate values fall in adjacent higher classes." Each of the first four classes is actually slightly smaller than the last. The difference in span, however, is very small and should create no problem. Such classes are usually considered to be of equal span. When, however, class spans vary in size by more than such a small amount, they are considered of unequal span.

When the values to be mapped have been organized into the desired number of classes, the original values are lost and the figures to be represented on the map will correspond to the class sequence numbers (in this case, 1, 2, 3, 4, and 5). Thus a limited number of integers is typically substituted for a much larger assortment of frequently complex and fractional figures.

The classes become the units of measurement in a very real sense. In terms of our example, we have classes 1 through 5 *in integers*. By means of the classing specification or the value key we know the range of values which each class embraces, but we no longer have any knowledge of the precise value applicable to each of the individual data zones.

When the classes are all of equal span, their employment as units of measurement presents no difficulties; but as we shall see later, when classes of unequal span are employed we must face the disquieting prospect, perhaps for the first time in our lives, of employing inconstant units of measurement. We expect a unit of measurement to be of fixed size. We are used to converting units of measurement from one system to another (feet to meters). We are used to units of measurement changing over time (the value of the dollar). We are even used to variation in the length of a degree of longitude depending on its latitude. But units of measurement that vary arbitrarily at one time and place present special problems.

With an unclassed map, full detail is provided; with a classed map, detail is sacrificed but generalization is achieved. Each approach has a different advantage. However, there is more to weigh than the merits of detail without generalization and the merits of generalization without detail.

WHY CLASS?

There are three possible reasons for grouping values into classes. The first is to reduce the volume of detail work required of the mapmaker. The second is to improve overall comprehension through the use of fewer numbers and distinctions. For example, we think about people's ages by years, rather than also considering months, days, hours, minutes, and seconds. We have all found ourselves listening to organization treasurers reporting their balances to the penny, only to have our comprehension of the financial issues diminished.

A third and frequently controlling reason is the choice of symbolism. All symbolism may be classed. EXTENT symbolism is the only type that as a practical matter can be continuously varied and which therefore is suitable for use with unclassed values. Even this type should be used unclassed only when the number of locations is relatively small (as in Foursquare) and the symbols are fairly large, so that the user may estimate their size with precision.

Use of Extent Symbolism with Classed Values

Of the EXTENT symbolisms shown in Chapter 9, let us first consider circle symbolism. When classing is used, an immediate issue to be faced is how to establish the circle area representing each class. The best solution may be to make the circle area proportional to the classing sequence, rather than related to the span of values. A map prepared on this basis for five classes is shown in Figure 10–1. The area of the circle used to represent the second class is twice that for the first class, the area of the circle for the third class is three times that for the first class, and so on. Choosing the circle size for the first class follows experimentation with these proportions. A practical advantage for hand production is that only five different sizes of circles have to be computed and drawn, compared to the 100 different sizes required when the values are unclassed. In addition, there is no problem in representing zero and values near to zero, since they are included in the first class.

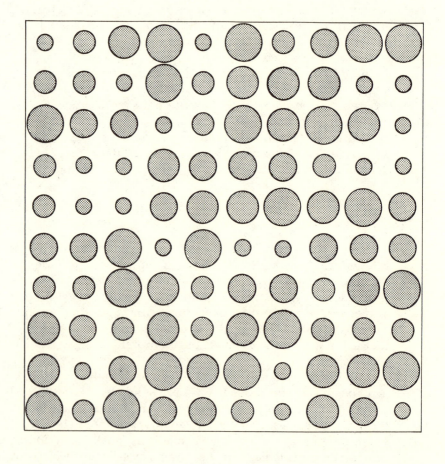

0–<20   20–<40   40–<60   60–<80   80–100

Figure 10–1   Spot-Extent: Circles, Five Classes (1-SE3)

DETERMINING SPOT SYMBOL SIZE (*Given Values Classed*)
1. Choose the smallest desirable or practical symbol that will still be noticeable.
2. Choose the largest desirable or practical symbol that will not create such a large range as to overpower the map.
3. Draw a straight line between them.
4. Divide off the scale line with the number of classes.

Note that the use of *both* end symbols is crucial in establishing the scale. If only the smallest or the largest is used as a base, there is no way to determine adequate differentiability.

When the minimum value of a range is a large percentage of the maximum value (92–100), circle sizes can be varied just as in Figure 10–1, though that value range is wider (0–100). Only the class spans are different.

With bar symbolism, a corresponding procedure is used. Bars for the fifth class will be five times as high as those for the first class. With a value range of 0–100, the map is more generalized but probably no easier to interpret, and there is no significant advantage to classing in terms of hand production.

With the raised conformant stepped symbolism shown in Figure 9–7, classing can be used with the result illustrated in Figure 10–2A. Again, the classing sequence numbers establish the various heights employed, and the display is considerably simplified. Instead of 100 separate stepped areas to be judged, there are about 60, standardized at five different heights.

Figure 10–2B shows a still more generalized display using only three classes. Here the number of separate stepped areas comes to fewer than 40, standardized at only three different heights.

To illustrate classing with sectors, we can look at Figure 9–10. For values which are percentages, it would be desirable to use either six or twelve classes, so that the arcs could correspond to five- or ten-minute intervals on a clock face. However, doing so would produce class spans defined by odd fractional values. It would probably be better to use ten or twenty classes, and to provide tick marks around the circle at six- or twelve-minute intervals.

When interpolation is employed, values should not be classed until after the interpolation process has been carried out, as prior classing would not only involve an extra operation but would produce less accurate results.

HOW MANY CLASSES?

We must now turn our consideration to the number of classes best employed. There is no one best number; the answer will necessarily vary with the circumstances.

In light of the three reasons given for classing and the restrictions inherent in successful symbolization, it is evident that a limited number of classes will prove most useful. The smaller the number of classes that can meet the

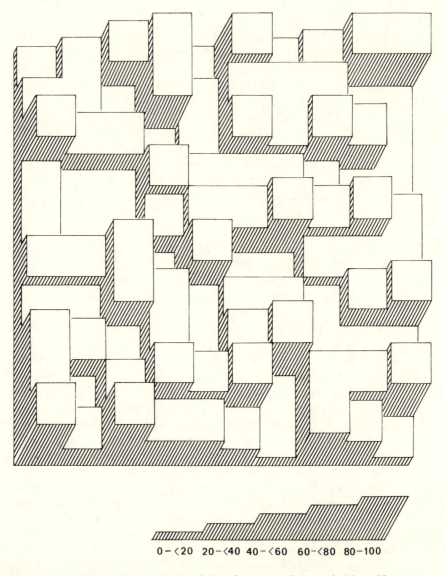

0–<20   20–<40   40–<60   60–<80   80–100

Figure 10–2A   Field-Extent: Raised Conformant, Stepped, Five Classes (7-FE4)

objectives, the greater the choice of symbolisms, the less work for the map-maker, and the greater the ease of comprehension for the user.

One can often intuit the number of classes on first consideration. If the data limits are respected and the class spans are equal, the limits of the class spans will be determined precisely by the number of classes selected. The mapmaker will often consider both aspects simultaneously. But, in any case, the number of classes chosen involves a subjective judgment and is likely to be a very important factor in the selection of symbolism.

Let us now come to grips with specific numbers, in terms of their functions

0–<33   33–<67   67–100

Figure 10–2B   Field-Extent: Raised Conformant, Stepped, Three Classes (7-FE4)

and problems. For some purposes, only two classes might provide the ideal solution—for example, territory under friendly control and territory under enemy control, or soil for growing watermelons and all other soil within the study area. With two classes, generalization is carried to its extreme. The world is white or black, good or bad, high or low, with nothing in between. In making any such differentiations, however, there is likely to be a "gray area" between the two classes.

A classic example of this problem is the differentiation of water from land. The definition of water must be clearly established. For a large-scale map of

a relatively small area such as a building site or the area of a microclima-
tological investigation, water might be defined to include any wet area large
enough to concern potential users of the survey. Thus, even a small spot,
only periodically wet, might be indicated by the thoughtful mapmaker. At
the other extreme, for maps of large regions, the test may be whether the
water areas are large enough to be represented at the scale of the map.
Finland and Canada both contain large regions full of lakes (not to mention
swamps and bogs), a fact likely to be quite interesting to map users concerned
with those regions. Yet because of the practical limitations of drafting, only
the largest of those water areas are likely to be represented. Some standard
or criterion for categorization must be established, whether rigid or flexible.

Dividing-line problems always arise when classing is to be employed. The
instance of two classes presents the problem in its most acute form. With
ten classes, a misplaced value falling in Class 8 is likely at most to shift up
to Class 9 or down to Class 7. With only two classes, any error in classing
would shift the value (and its location) from the highest to the lowest class
or vice versa, but misclassification is less likely.

With three classes the problem is immediately alleviated, as the middle
class can always represent the borderline situations: areas of uncertain military control, pretty good but not ideal watermelon soil, or swampy land. If
two represents the minimum number of classes, what is the maximum? No
more than ten classes are usually employed, but is there an *optimum* number
of classes most likely to be useful?

The answer is almost certainly five. Dividing data into five classes provides
a central, a high and a low, and an extra high and extra low class. The central
class serves as a benchmark in relation to which the other classes can be
considered, contributing importantly to comprehension. When a middle class
is deemed useful but more detail is desired than can be provided by the use
of five classes, seven classes is a sensible choice. Eight classes might be
employed when no middle class but more detail or greater differentiation is
desired. With eight classes, however, the point of optimum compromise
between detail and generalization may have been exceeded. Aside from the
problems of symbolization, it is no longer possible to relate to nearby points
of reference. In view of the relative difficulty of categorizing and differentiating the six middle classes, the highest and lowest classes may receive too
much attention; or the four central classes may become invisible. Nine classes
are seldom employed.

In a culture accustomed to the decimal system, the use of ten classes has
a certain logical appeal and may at times be warranted. Experience suggests,
however, that under most circumstances greater comprehension may be
gained by the use of five or six. This may be the case because it is more
difficult for the map user to differentiate adjacent levels of symbolism. Sometimes, however, the symbolism may establish larger categories than the
individual classes themselves.

COUNT symbolism is usually most effective with ten classes. With some
COUNT symbolisms a larger number of classes can be used successfully.
The outside limit, however, should be set at 25 classes. For COUNT sym-
bolism of the SPOT type, the optimum number of classes is probably nine

or sixteen, with the symbol elements arranged in a square pattern, three or four elements on a side. When EXTENT symbolism is classed, a maximum of five or six classes is ideal. To attempt more detail may actually result in less communication.

With DARKNESS symbolism, as noted, ten classes is the absolute upper limit without resort to color. Five or six classes will be the usual limit with gray patterns based on typical dot screens, employing high quality printing if the map is reproduced.

The optimum number of classes is thus limited by constraints specific to the type of symbolism chosen.

VALUE POSITIONS AND VALUE CURVES

When considering classing, the positions of the values within the study space and within the value range are very important. In our Foursquare example, the values might have been located differently within the study space. Or when arranged sequentially by increasing magnitude, instead of being more or less uniformly distributed within the value range, they might have been unequally distributed in an infinite variety of ways. The construction of a value curve (Figure 10–3) will prove helpful in understanding the distribution of the values within the range.

The first step in constructing a value curve is to rank-order the data values from lowest to highest (Figure 10–3, large upper diagram). This step is very important and precedes *any* classing manipulations. A square-format graph is constructed with a point for each location's value plotted in equal spacing from left to right; the lowest value is placed at the lower left corner, and the highest value at the upper right corner. An engineer's scale will aid in achieving equal spacing for any number of values.

The vertical position of each point is determined by its value relative to the value range. In the case of Figure 10–3 this is not difficult to grasp, as the value begins with 0.

The value curve for our Foursquare example resembles the first of the small lower diagrams in Figure 10–3, the straight line. In real life, this is the rarest distribution encountered.

A value curve may actually be of any shape, as long as it starts at the lower left corner, moves continuously toward the right without at any time proceeding at an angle less than horizontal, and ends at the upper right corner. For the purposes of establishing the basic nature of a value set, however, value curves may be usefully assigned to one of the five categories illustrated:

1. Those that are more or less straight, having a linear rate of increase (constant slope);
2. Those that lie to a significant degree toward the lower right corner, having an increasing rate of increase (low, increasing slope);
3. Those that lie correspondingly toward the upper left corner, having a decreasing rate of increase (high, decreasing slope);
4. Those that have a significant shift after starting in the low position (low, followed by high);

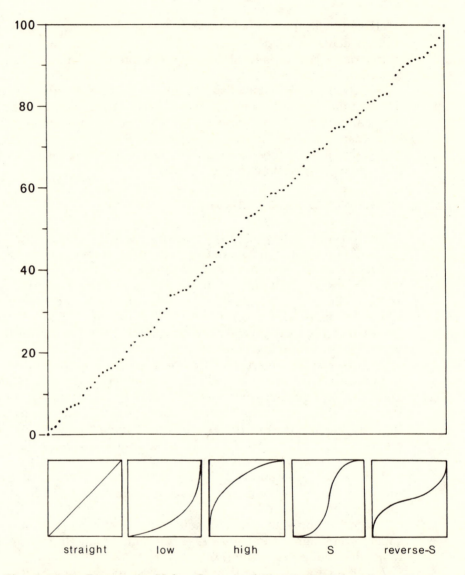

Figure 10–3   Foursquare Value Curve and Five Basic Value Curve Types

    5. Those that have a significant shift after starting in the high position (high followed by low).

Let us illustrate the significance of these two variables of position and value by considering six possible alternatives that might have been encountered in mapping Foursquare.

Alternative 1A: Straight Value Curve, Random Positions   Assume a slightly different set of values (Table 10–1) yielding a *precisely* straight value curve like that shown in the first of the small lower diagrams of Figure 10–3. The value range will be 0–99, increasing from zero in even digital steps (0, 1, 2, . . . , 98, 99). For these values distributed randomly among the locations,

Table 10–1: Foursquare Values—Alternative 1A

Locations	Values	Locations	Values	Locations	Values
1	1	35	24	68	11
2	61	36	77	69	62
3	25	37	92	70	42
4	52	38	31	71	29
5	16	39	4	72	90
6	8	40	15	73	81
7	5	41	22	74	46
8	33	42	93	75	73
9	19	43	45	76	97
10	91	44	28	77	94
11	87	45	21	78	78
12	18	46	23	79	85
13	47	47	13	80	86
14	41	48	98	81	51
15	82	49	49	82	89
16	14	50	80	83	34
17	60	51	56	84	53
18	59	52	79	85	38
19	17	53	68	86	43
20	26	54	12	87	66
21	40	55	74	88	96
22	76	56	64	89	2
23	99 max.	57	83	90	88
24	75	58	57	91	69
25	72	59	95	92	10
26	67	60	71	93	39
27	30	61	48	94	50
28	84	62	20	95	37
29	32	63	36	96	3
30	6	64	55	97	0 min.
31	44	65	27	98	63
32	7	66	65	99	58
33	35	67	54	100	70
34	9				

the resulting maps are shown in Figures 10–4A and B. Any or all of the other symbolisms so far illustrated might have been employed, but these two in combination are particularly revealing. The conceptual surface, though similar in type to that previously shown in Figures 9–14B and 9–15F, is radically different in terms of positioning of the values involved. By chance, the corresponding extreme values happen to fall in almost the same positions, but otherwise there is little or no similarity. Note, for example, the internal pit to be seen near the lower right corner of the new surface— nothing comparable has previously been encountered.

Alternative 1B: Straight Value Curve, Ordered Positions   In Figures 10–5A and 10–5B we show a more startling demonstration of the significance of position

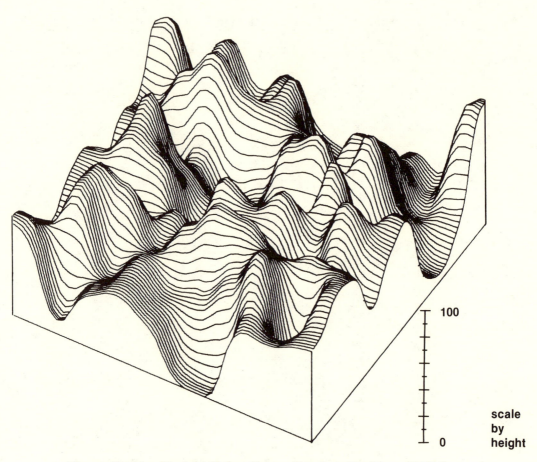

Figure 10–4A   Straight Value Curve, Random Positions (7-FE6)

alone. The values are identical to those just shown, but spatially arranged in order of increasing magnitude starting from the lower right corner. The spatial disposition here yields the smoothest conceptual surface possible with the assumed values. If we employed the value set originally assumed for Foursquare in these two maps, the resulting conceptual surface would have been quite similar but slightly irregular.

Alternative 2A: Low Value Curve, Random Positions   Assume a low value curve with the positioning of the values resulting in Figures 10–6A and 10–6B. While the topmost point of the highest peak is exactly the same height as before, the rest of the surface has slumped—just as the low value curve appears to be a drooping version of the straight value curve. There is a far greater prevalence of low values, and a lower incidence of high values. In consequence, a much larger portion of the plane interpolated map (Figure 10–6B) falls in the first class, more than half of the entire study area. In contrast, less than 1 percent of the entire area of the map falls in the fifth class.

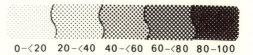

0–‹20   20–‹40   40–‹60   60–‹80   80–100

Figure 10–4B   Straight Value Curve, Random Positions (8-FD7)

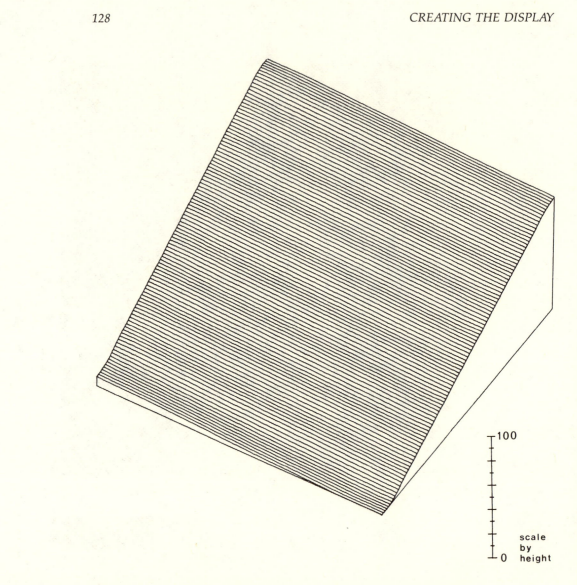

Figure 10–5A   Straight Value Curve, Ordered Positions (7-FE6)

0–‹20   20–‹40   40–‹60   60–‹80   80–100

Figure 10–5B   Straight Value Curve, Ordered Positions (8-FD7)

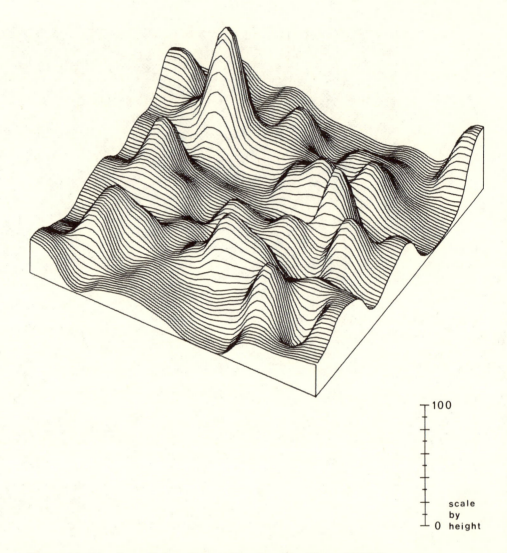

100
scale
by
0 height

Figure 10–6A   Low Value Curve, Random Positions (7-FE6)

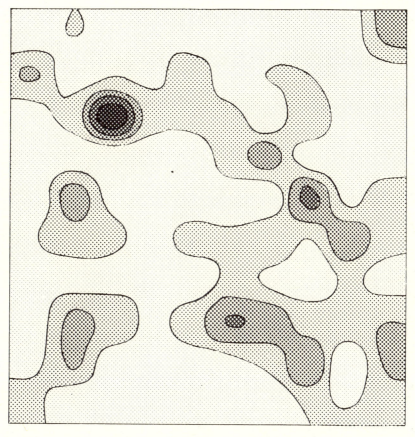

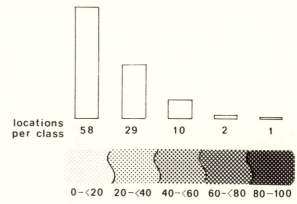

Figure 10–6B   Low Value Curve, Random Positions (8-FD7)

Figure 10–6C shows the same facts using plane conformant symbolism. Note that reading this map is considerably more difficult.

Figures 10–6D and E show the use of dot symbolism with the locations both differentiated and undifferentiated. Note that because of the large num-

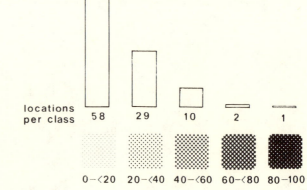

Figure 10–6C   Low Value Curve, Random Positions (8-FD4)

ber of classes involved compared with the preceding two maps, much greater differentiation is provided within those portions characterized by low values. Here no class represents more than 14 percent of the map area as compared with 58 percent in the first class of the plane conformant map in Figure 10–6C.

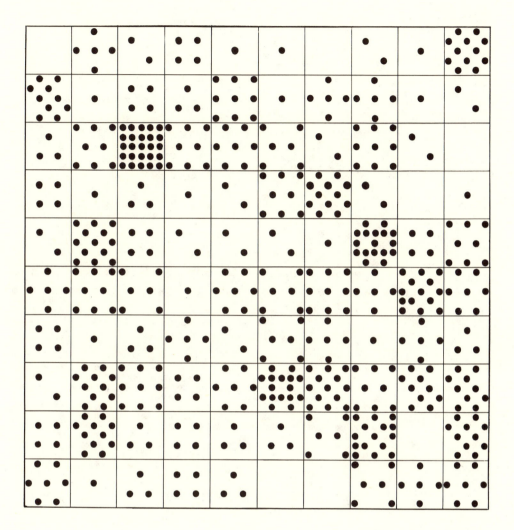

● = 4
not shown for < 2

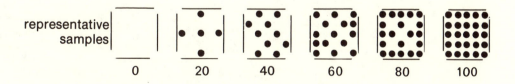

representative
samples

0   20   40   60   80   100

Figure 10–6D   Low Value Curve, Random Positions, with Location Boundaries (8-FD5)

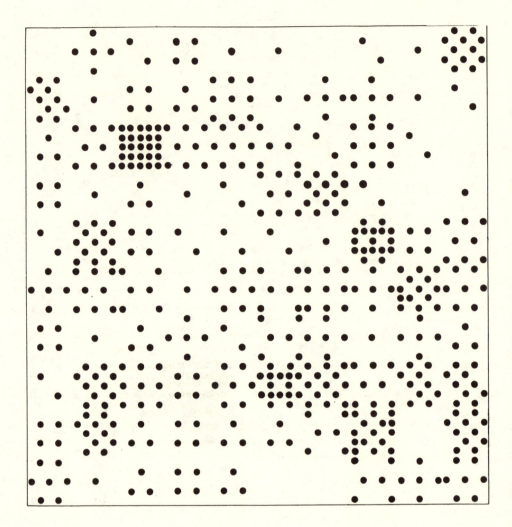

● = 4

not shown for < 2

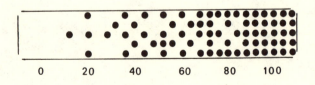

Figure 10–6E   Low Value Curve, Random Positions, without Location Boundaries (8-FD5)

Alternative 2B: Low Value Curve, Ordered Positions   In Figures 10–7A and 10–7B, as in the preceding alternative, more than half the entire area of the map falls in the first class, with less than 1 percent in the highest or fifth class. Figure 10–7C shows the same facts mapped by the use of plane conformant symbolism. Figures 10–7D and E again show the use of dot symbolism.

Alternative 3A: High Value Curve, Random Positions   In Figures 10–8A and 10–8B, the surface appears to have risen. Over half the map falls in the fifth or highest class, 87 percent of the locations fall in the highest two classes, and less than 1 percent fall in the lowest class.

Alternative 3B: High Value Curve, Ordered Positions   The result of positioning the values sequentially is shown in Figures 10–9A and 10–9B. If the conceptual surface of Figure 10–7A were turned over diagonally, it would mate

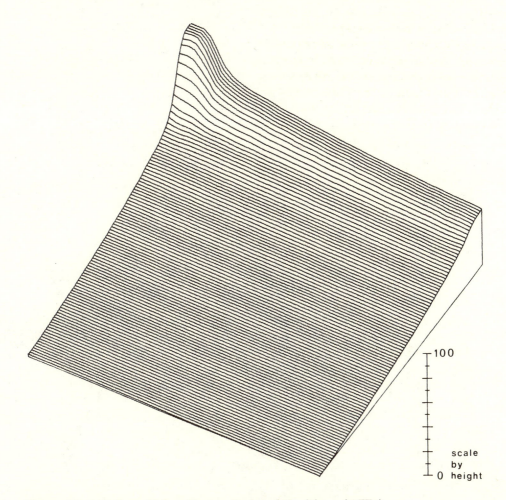

100

scale
by
0   height

Figure 10–7A   Low Value Curve, Ordered Positions (7-FE6)

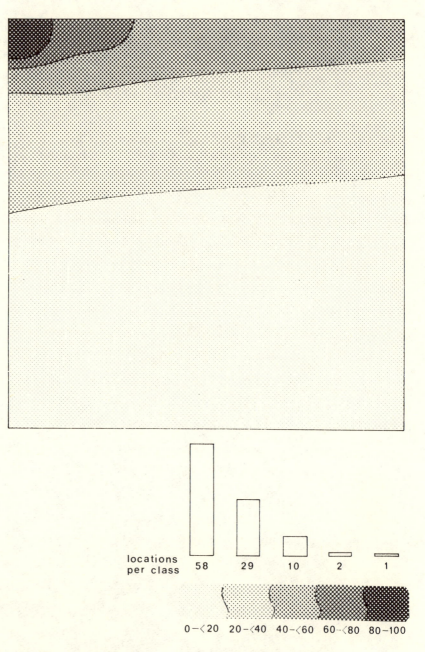

Figure 10–7B   Low Value Curve, Ordered Positions (8-FD7)

perfectly with Figure 10–9A. This conformance could have been anticipated from inspection of the value curves themselves, as shown in the second and third small diagrams of Figure 10–3, but only when the values are sequentially ordered.

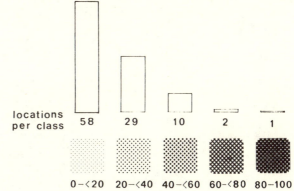

Figure 10–7C Low Value Curve, Ordered Positions (8-FD4)

S and Reverse-S Value Curves The results of the value curves of the two remaining types shown in the small diagrams of Figure 10–3 may be easily visualized. For the "S" type there would be many low and high values and few intermediate values. Thus with raised interpolated symbolism there would be numerous peaks and pits with little area of intermediate height

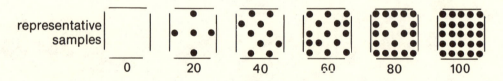

Figure 10–7D   Low Value Curve, Ordered Positions, with Location
Boundaries (8-FD5)

between, while with plane interpolated symbolism there would be mostly light and dark, with little area of intermediate gray. For the "reverse-S" type there would be few low and high but many intermediate values. Raised interpolated symbolism would show few peaks and pits, with much of in-

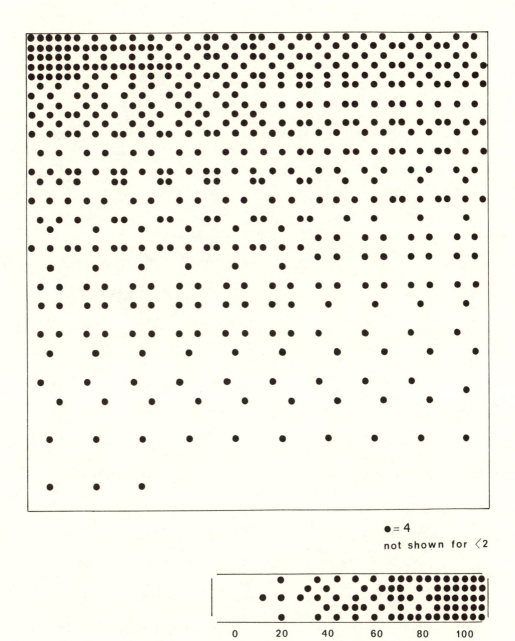

● = 4
not shown for ‹2

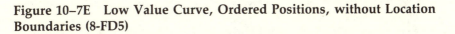

0   20   40   60   80   100

Figure 10–7E   Low Value Curve, Ordered Positions, without Location Boundaries (8-FD5)

termediate height between, while with plane interpolated symbolism there would be mostly intermediate gray.

Frequency of Curve Types

The statistician's bell-shaped curve of "normal frequency" produces a value curve of the reverse-S type. However, since a normal frequency curve ex-

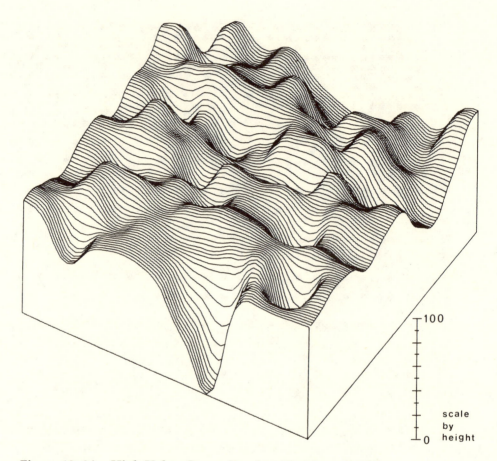

Figure 10–8A   High Value Curve, Random Positions (7-FE6)

tends to infinity left and right, it cannot be drawn as a value curve. It can be represented only by making certain arbitrary assumptions, the nature of which will determine the result.

Surprisingly, in terms of spatially variable data, the reverse-S or "normal frequency" type is rarely encountered in mapping practice. This is also true for the last two types considered, the "high" and "S" types. Low value curves are the most usual.

The subject of literacy in developed countries may be cited as an example of a high value curve. If the subject is adjusted to represent illiteracy, however, the curve becomes low. Since there are advantages in using low value curves, their prevalence no doubt results in part from such adjustments.

A map with many low values and few high values is usually easier to make and interpret than the converse. With oblique views of all kinds it is easier to interpret isolated peaks than partially hidden corresponding pits, but given a choice of symbolisms, it will most often be preferable to select a map title that can yield a low value curve. With circles and bars, for example, it is easier to differentiate among numerous small symbols than among large symbols, and with low values the map will be less cluttered.

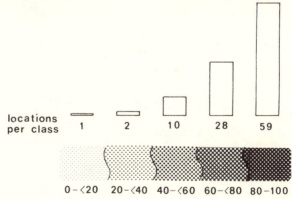

Figure 10–8B High Value Curve, Random Positions (8-FD7)

With darkness symbolism, information can be presented more effectively against lighter than against darker tones. For such reasons, a map showing houses in a cold region *without* central heat would be preferable to a map showing houses *with* central heat, but the latter subject would be preferred

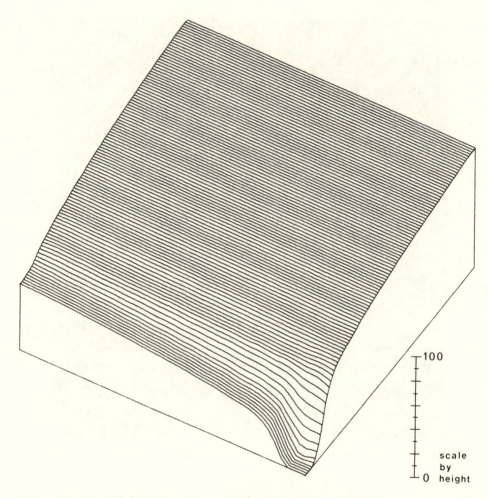

Figure 10–9A   High Value Curve, Ordered Positions (7-FE6)

in a warm region. In a region of intermediate climate, a situation corresponding more closely to the normal frequency distribution would be expected, resulting in the value curve of the reverse-S type.

Use of Roots and Powers

When value curves are especially low and raised symbolism of either conformant or interpolated type is employed unclassed, there will be little differentiation among the numerous smaller values. For example, if population
per square mile by counties for the coterminous United States is to be displayed, a few high peaks would appear at the major centers of population
while most of the rest of the country would be almost undifferentiated.
Though such a display would be quite accurate, a map showing greater
differentiation among the numerous lower values would be more revealing.

Such a map can easily be achieved by using the square roots of the given

Figure 10–9B   High Value Curve, Ordered Positions (8-FD7)

values.* By this simple procedure, the lower portions of the map become somewhat higher and variations present become more easily differentiated. (At the same time the higher portions become contracted, and variations

*The difficulty can also be resolved by the use of unequal classing.

there become less easily differentiated.) Among the middle values, differentiability will remain about the same.

Cube roots, fourth roots, or others—including fractional roots, such as the root 2.6—might equally well be used; the choice depends upon the nature of the value curve. In general, the lower the value curve, the higher the root rating likely to prove useful and the greater the differentiation provided among low values.

If the value curve is high, it is possible to employ exponents in a similar manner. If all the given values were squared, for example, the result would be more differentiability among the higher values and less among the lower values.

The map user must understand the nature of the rather special adjustment that is being made. However, since root and power progressions are inherently orderly, they may easily be grasped by the user if presented in the value key.

Figure 10–10 shows a family of root and power curves suitable for use, assuming that the values will be expressed in terms of percentage between the extremes and that a balance in differentiability is needed over the full range. The root curves are toward the lower right while the power curves are toward the upper left. The best curve may be selected by comparing the applicable value curve with the root and power curves shown in the chart.

To illustrate, let us assume that more differentiation is desired among the lower values of the raised interpolated display shown in Figure 10–7A. By comparing the low value curve of Alternative 2B (shown in Figure 10–16A) with the root curves of Figure 10–10, we can see that cube roots might be appropriate.

If the fit between the value curve and the selected root or power curve happened in any particular instance to be perfect, differentiability would be equal over the full value range. The straight curve at 45° provides equal differentiation throughout the value range. Thus the point at which one of the other curves is parallel or at 45° is the point of transition between greater and less differentiation. The two transverse lines drawn on Figure 10–10 show the approximate position on the root and power curves of their points of tangency to a line parallel to the 45° slope of the straight curve. All locations which fall to the left of the upper line will be more differentiated, and those which fall to the right will be less differentiated. Locations falling near these lines will be affected only slightly.

The point of transition for the cube power curve (3) lies further to the left than the point of transition for the square root curve (2). To the left of these points the curve is steeper than the straight curve, providing greater differentiation; to the right the curve is more horizontal than the straight curve, providing less differentiation. Hence the *percentage* of the values at the low end with greater differentiation is smaller with cube powers than with square roots (Figure 10–11). To comprehend the relative power of root curves, it is only necessary to glance at the corresponding power curve and note approximately where it is crossed by the transverse line.

With oblique views based on S and reverse-S value curves, it is possible to combine the root and power approaches. S value curves would receive

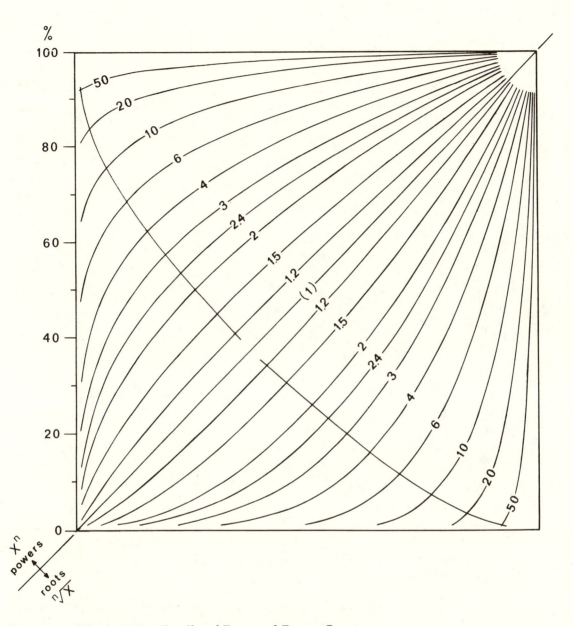

Figure 10–10 Family of Root and Power Curves

greater differentiation toward the flatter end portions, and reverse-S value curves would be differentiated more in the flatter middle portions.

If the minimum given value is a moderate-to-large percentage of the maximum, the value range should be converted to a 0–100 percent range before applying roots or powers. When preparing the value scale that appears on the map, the value range must be converted back to the original numbering system.

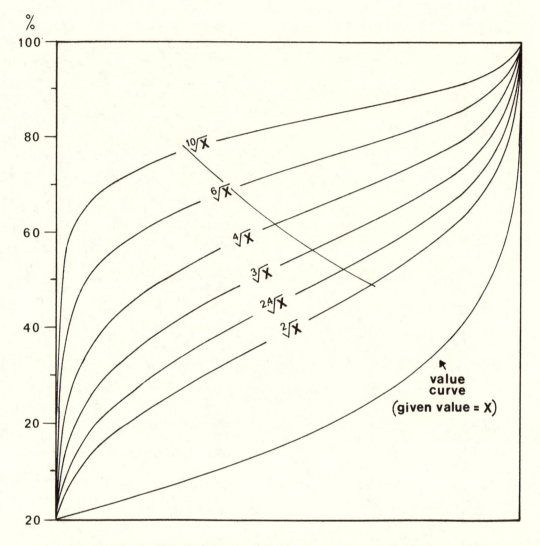

Figure 10–11   Effect of Roots and Powers on Low Value Curves, Alternative 2B

   The root and power procedure may be applied to any type of unclassed symbolism—though this is rarely necessary, since each separate value is revealed. For example, the length of bars or the area of circles might be varied as was done with the heights of raised conformant and interpolated surfaces, though circle symbolism, as was previously mentioned, inherently facilitates comparison among smaller values. The procedure may also be used with plane interpolated symbolism, which is unclassed only in part (where the different tones meet). Application of roots or powers to SPOT symbolism is usually confusing at best.
   Value curves can be very useful to a designer, but they should not be misapplied. Assume a study space consisting of a single row of ten locations, with the values shown below each location: if the values were spatially

ordered according to increasing magnitude, they would be easy to differentiate throughout. Adjusting them according to a suitably chosen root curve makes them even easier to differentiate.

Thus, if the sole objective is to compare the relative heights applicable to the smaller values, using roots will clearly be helpful. If, however, the objective is to understand the trend as a whole, the use of roots will seriously confuse the situation. The representation of the true facts in the profile at the lower left is far better for most purposes than the adjusted profile at the lower right.

Figure 10–11 charts the varying effects of roots and powers on the low value curve. The higher curves show the trends that would result if the specified roots of all of the given values were taken and then readjusted. To illustrate the process, let us assume cube roots are being considered. The cube root of each given value is obtained and the resulting figures are then multiplied by 21.5, which is the number of times the cube root of the maximum value (100) must be multiplied to regain that value. The results are plotted as dots, and a curve is drawn through them. Since we are usually concerned with the trend of the new curve rather than its precise delineation, it is adequate to select a few representative values across the width of the chart and draw the line with a French curve. In the present instance, values were plotted at locations corresponding to even multiples of ten.

Among the various curves shown in Figure 10–11, that for root 2.4 most nearly approaches a straight line curve, and hence it would yield the best differentiation over the full value range.* The effects to be anticipated can be broadly judged by comparing the slope of each curve across the chart with the slope of the original value curve. Wherever a portion of a new curve is steeper than that portion of the original curve lying directly below it, the new curve will yield increased differentiation, and vice versa. The transverse line shown represents the approximate points of transition between the two situations.

EQUAL CLASSING AND SUBCLASSING

When classing is to be employed, classes of equal span are greatly desirable whenever possible. Classes are units of measurement in a very real sense, and we expect units of measurement to be of constant size.

Classed darkness symbolism is frequently used for plane interpolated and plane conformant maps. An unfortunate result may be the construction of large concentrated areas without differentiation. Yet the displays of Figures 10–7B and 10–7C are accurate, and highly revealing in that they show exactly where the smaller values lie. If the use of equal classes is ever justified, it is certainly justified in the case of these two maps. These maps are exceptions to the general rule that when a relatively large percentage of a map is un-

---

*If only slight differentiability is desired among lower values, roots less than a square root, such as root 1.5, might be used.

differentiated—and especially when that percentage is concentrated in a few areas—classed darkness symbolism should probably not be used.

With the plane interpolated symbolism of Figure 10–7B, it is easy to provide greater differentiation within the first class or two through the use of equal subclasses. Such a procedure is like resorting to quart measurement when measurement by gallons proves too gross. Figure 10–12 shows such a solution. Each of the first two classes has been subdivided into four equal classes by means of dashed contour lines. In each case the increasing magnitude of the values is indicated by the increasing lengths of the dashes used. By this means the area and the number of locations in each class has been more nearly equalized. Whereas there were formerly as many as 58 locations in one class, the maximum is now 18—less than 20 percent of the total number of classes—with a corresponding increase in the information that can be communicated over the map as a whole.

The same procedure might be used for the plane interpolated map of Figure 10–6B (Alternative 2A, the low value curve with random positions). However, because the area falling in the first class or two is far less concentrated in that map, the need for such action is far less urgent. If equal subclassing were to be employed for such a small map, it would be desirable to avoid overcrowding by limiting its use to the first class (Figure 10–13). With a larger map, additional equal subclassing might be employed successfully if warranted.

In the sense that equal subclassing involves classes of two different spans, it is of course unequal. However, since the spans of the subclasses are all the same and represent an equal division of the original equal class spans, it can be thought of as a variation of equal classing.

It is not possible to employ equal subclassing with plane conformant symbolism because each location is treated entirely independently. Differentiation is difficult between equal classes and equal subclasses without the aid of color. Given the inherent virtues of equal classing, we might simply accept the result which it yields. With an extremely low value curve, however, this may not prove useful. Unless the designer is prepared to shift to the use of some other symbolism, the only recourse is to alter the classing scheme (i.e., to use unequal classing).

ROUNDING

Rounding, the practice of expressing numbers in simpler terms than given, is sometimes troublesome. Individual values are best used by the map designer as given. Our main concern is therefore class limits and extreme values.

Rounding is relative. The magnitude of the value range can be increased but should never be decreased. Rounding a value range of 1–99 to 0–100 is trivial, but rounding a value range of 1190–1825 to 1000–2000 is drastic. Rounding that is other than negligible is not recommended.

With extreme values of 21 and 82 and the wish to use five equal classes, for example, the unrounded solution would be to divide the difference be-

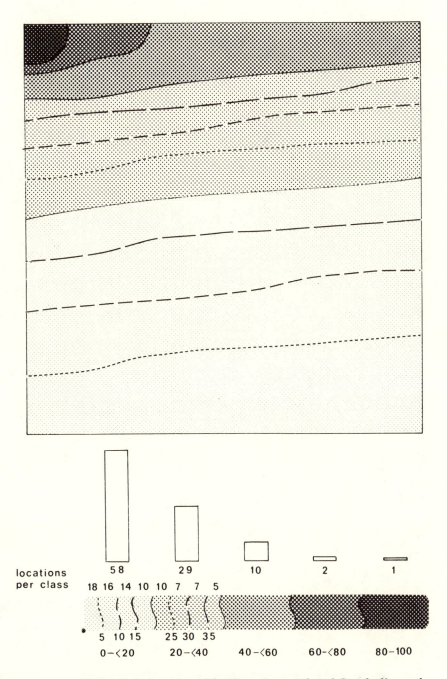

Figure 10–12   Equal Subclassing with Plane Interpolated Symbolism of Figure 10–7B   (Low Value Curve, Ordered Positions) (8-FD7)

tween the two extreme values (61) by five. Each class span would thus equal 12.2. This figure could not be rounded down to 12, as the range would then be smaller than required. Rounding up to 13 would lead to the range of 20–85, perhaps satisfactory. Looking at the class spans, however, we find that

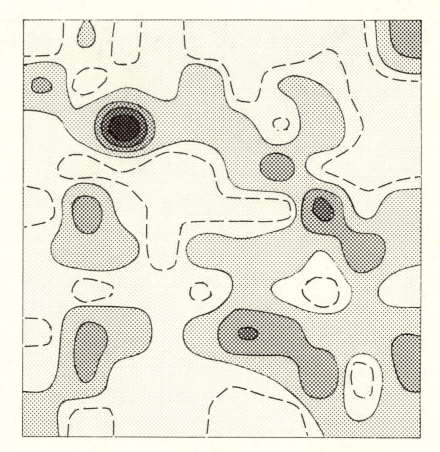

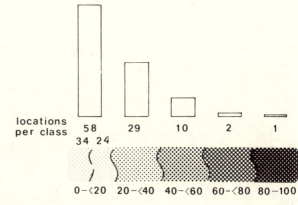

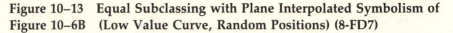

Figure 10–13   Equal Subclassing with Plane Interpolated Symbolism of Figure 10–6B   (Low Value Curve, Random Positions) (8-FD7)

they would be 20–<33, 33–<46, 46<59, 59–<72, and 72–85—which, in view of the odd integers involved, is less than "well-rounded." A span of 15 for each class is probably the minimum acceptable. This would make the value range 15–90, with class spans of 15–<30, 30–<45, 45–<60, 60–<75, and 75–90.

Such rounding always raises problems in the two extreme classes. In the last example, out of the total range for the first class of 15–<30, only 21–<30, or 60 percent, is being used. The effective span of the last class is only 75–82, or 47 percent. In consequence, the stated range of 15–90 would be highly misleading, since map users inevitably assume that all or most of the span of an extreme class is being used. This is a case where a detailed value key would be advantageous.

UNEQUAL CLASSING

With unequal classing, class spans are varied in such a way as to divide the number of locations—and if the locations are similar in size, the area of the map—more equally among the various classes and their symbolisms. Unfortunately, even under the best of circumstances the use of unequal classing imposes a special burden upon the map user. At the worst, when class spans appear to be arbitrary, prolonged study may be required of the user to interpret a map correctly.

The presence of unequal classes is seldom revealed graphically, and a good value key graphically showing the value range per class becomes critical. The map user must remember the values applicable to each varying class span, at least approximately, or the map will be misinterpreted. Yet under normal circumstances it is almost impossible for even skilled map readers to remember the values occurring at class breakpoints while examining the map. Equal classing gives the map meaning when the user observes and remembers the extreme values used in the overall classing range. When unequal classes are employed, however, the value at each breakpoint assumes importance. With six classes, for example, there would be five breakpoints and seven values to remember.

To minimize this difficulty, the best procedure is some system by which the class spans may be varied *logically*—and hence comprehended more easily than the actual breakpoint values for the unequal classes. When such logic is used and graphically communicated, the special burden imposed by unequal classing is greatly reduced, though not entirely eliminated.

Quantile Classing

One system which is frequently employed is the substitution for classes of equal span of classes containing an equal number of locations when arranged according to increasing magnitude. For example, if the number of locations is divisible by four (such as thirty-two), four classes each representing eight locations would be used. By this method, identified by the general term *quantile*, class spans may be highly irregular, but this becomes a secondary consideration. The major emphasis is placed upon the equality of location number in each class. Thus, when four classes are employed, the lightest tone will represent that fourth of the locations with the smallest values, the next-to-lightest tone will represent that fourth of the locations with the next-to-smallest values, and so on. Quantile classing by four classes is called quartile, by five, quintile, by six, sextile, and by ten, decile.

This procedure is often quite successful. It provides an easily understood logic and offers the great virtue of applicability regardless of the shape of the value curve. Two difficulties, however, are frequently encountered in proposed applications. Together, they limit the general use of the quantile system, especially with maps in series for purposes of comparison. If the same value falls on each side of a desired breakpoint, it may be difficult to place the same number of locations in each class. Two or more locations of identical value obviously cannot be assigned to different classes and symbolized in different ways. Similarly, if the number of locations is not evenly divisible by any desired number of classes, precise quantile classing will be impossible. However, since the difference is never more than one digit, in a map with a large number of locations, the resulting inequality will be slight and may not be objectionable. Under such circumstances, the lightest tone, for example, would represent *approximately* rather than precisely that fourth of the locations with the smallest values. (When a slight inequality exists that does not rule out the quantile approach, the mapmaker must choose by logic or hazard which class or classes to make smaller than the others.)

The number of locations of identical value straddling a class breakpoint may be substantial, especially if the given values have been recorded or prematurely rounded to less than three or four significant figures. If this is the case, one might consider altering the number of classes and defining new breakpoints that assist equality of observations per class. Otherwise, the number of locations in each class would become so divergent that the logic of the system would be destroyed.

With certain subjects that have a considerable number of small locations, the value curve may begin with numerous locations with a value of zero. If the locations of zero value extend beyond the first class breakpoint, the problem may frequently be resolved by changing the map title to exclude them. For example, for a map showing the number of dilapidated houses per city block, the title might be worded either "Presence of Dilapidated Houses by Blocks" or "Dilapidated Houses by Blocks Containing One or More." Blocks with a value of zero would then become interspace within the study space.

Figure 10–14 shows a quintile map of plane conformant type for the low value curve and ordered positions of Alternative 2B. The value key gives the number of locations per class, and reveals the great inequality in the class spans. The lightest tone now represents that 20 percent of the locations with the lowest values, those of 0–<5.60. The quantile approach is reasonably successful even though the four breakpoints between the classes are extremely difficult to remember. Since the locations are of equal size, there is an equal area of tone for each of the five classes.

Because the low value curve in this example is so regular in form, the progression in the "value span per class" is necessarily regular. However, when the quantile system is used with an irregular value curve, the size of the class spans may be equally irregular. Large and small class spans may alternate or otherwise intermix when proceeding from low to high values. The class spans will not usually progress logically when the value curve has even small reversals.

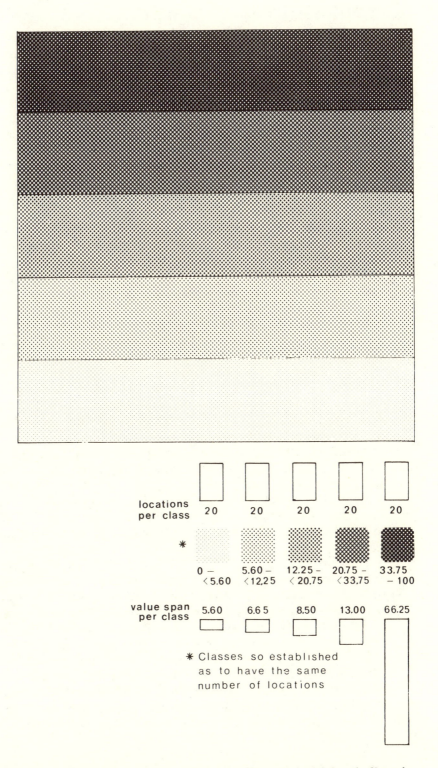

Figure 10–14 Quintile Classing with Plane Interpolated Symbolism for Alternative 2B (Low Value Curve, Ordered Positions) (8-FD7)

If we had illustrated Alternate 1B (the straight value curve, ordered positions) by a plane conformant map, it would be identical to Figure 10–14 except for the value key. Alternative 2B (the low value curve, ordered positions), illustrated by a quintile map of plane interpolated type, would be similar to Figure 10–5B except for the value key. These observations stress the inherent problem with maps based on unequal classing of any type—namely, that their appearance is no clue to what they represent. The display itself is only crudely indicative of the facts being presented. With value sets as different as those just cited, the maps should appear equally different—and certainly they should not look the same. Values are seldom as regular in space as our Foursquare examples, but spatial regularity is commonly encountered in mapping work. Histograms marked with class boundaries should accompany the value key, to provide information about frequency distributions and facilitate comparison of maps with different distributions.

Reciprocal Curve Classing*

Classing by reciprocal curves may be used when the value curve is low or high, but the values within the range must be converted to percents. Reciprocal curve classing employs classes that increase in span by a smooth logical progression. Any number of classes may be used, and the class spans in any given instance depend upon the nature of the value curve.

When class span sizes are plotted for low value curves, the classing curve will start low for the first class (the smallest span), get increasingly higher, and end high for the last class (the largest span). There will be no reversals as no two classes can have the same span. This progression is usually easily comprehended and remembered.

Arithmetic and geometric (including logarithmic) curves have limitations that make them unsuitable for general applications: some depend upon specific maximum and minimum values, and most are asymmetrical and present other problems. The beauty of the reciprocal classing approach is that a family of symmetrical curves (Figure 10–15) can be used with a table to compute breakpoints for any high or low value curve for any number of classes.

The chart of Figure 10–15 shows representative examples from the infinity of possible curves of this type. Curves designated "I" followed by a number (upper left section of chart) are inverted, and used for classing high value curves. The mapmaker plots the value curve, lays it over the chart to discover the closest reciprocal curve, and references it in the table, as described in Appendix 1. Choosing among curves is a simple matter, a subjective yet acceptable procedure. The most important goal in establishing unequal classes is an easily understandable flow of class spans. Compared with that, determining the one best possible curve is far less significant.

For the method of using reciprocal curve classing, see Appendix 1.

Look now at Figures 10–16A, B and C. Maps B and C show respectively,

*The reciprocal classing system was developed by the author.

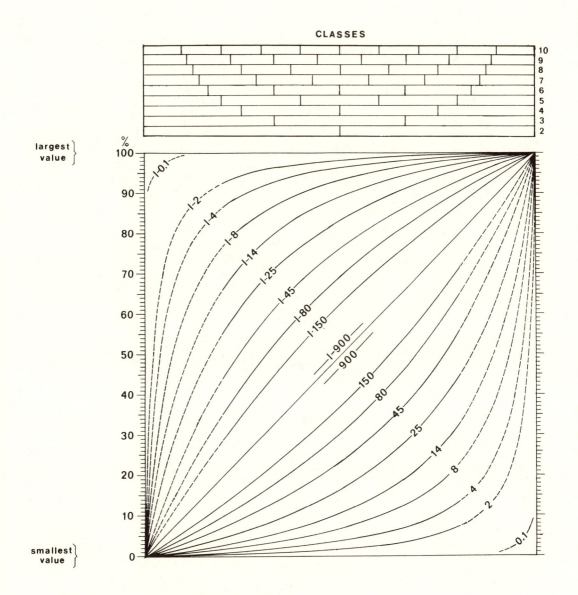

Figure 10–15 Reciprocal Classing Curves

for the low value curve of Alternative 2B, the use of plane interpolated and plane conformant symbolisms with the reciprocal curve system for unequal classing. Their value keys show not only the number of locations per class but also the value span per class. The latter illustrates the systematic and logically progressive increase in the class spans—easily grasped by the map user and retained at least approximately as the map is examined. The differences among the upper bars reflect the lack of conformance between the value curve and the class curve. There is no reason to expect the two curves to agree precisely, but it is desirable to have as nearly the same number of locations in each class as is practical.

Charting Value Curves and Classing Curves Together   The chart of Figure 10–16A illustrates the relationships between the value curve and the classing curve, and shows how in combination they determine the value span per class and the number of locations per class. The value curve has been plotted by the procedure previously described. If plotted on a much larger scale it would reveal small local irregularities. (Since it was earlier decided that the bottom class limit should be rounded down to 0.0, the dot representing the actual given minimum values of 0.3 correctly appears near the lower left corner, but slightly above the zero value line. If rounding had not been used, the actual minimum value would have appeared precisely at the corner.)

The classing curve has been plotted according to reciprocal curve 20 from Appendix 1. The nine dots were drawn and then connected with the aid of a French curve. Proceeding from left to right, the first dot was placed 1/10th

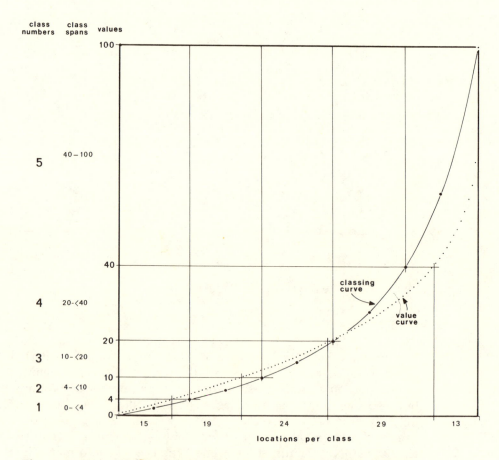

Figure 10–16A   Value and Classing Curves, Reciprocal Curve 20

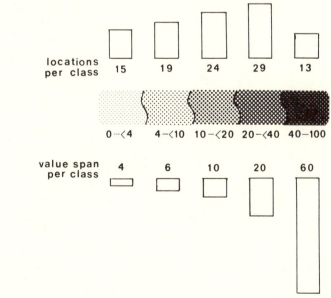

Figure 10–16B   Plane Interpolated Symbolism, Reciprocal Curve Classing, Alternative 2B (Low Value Curve, Ordered Positions)

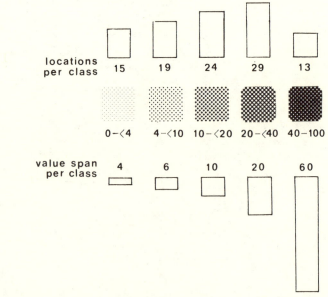

Figure 10–16C   Plane Conformant Symbolism, Reciprocal Curve Classing, Alternative 2B (Low Value Curve, Ordered Positions)

of the distance across the chart (as shown by the first of the upper thin vertical lines) and at a height above the zero value line of 1.82, the first figure from the table. The second dot was then similarly placed 2/10 of the distance across at a height of 4.00 (the second figure from the table), and so on.

Horizontal lines were then drawn across the chart at the breakpoint values for the desired five classes. The respective distances between these, including the top and bottom chart borders, are in direct proportion to the value spans per class. These distances also correspond to the heights of the lower bars of Figures 10–16B and 10–16C.

Classing Relative to a Medial Base

In circumstances when equal classing appears inappropriate and the unequal classing methods so far considered are unsatisfactory, it may be useful to consider classing in relation to what we will call a *medial base*. With this method, the class spans are arranged in an independent series on each side of some figure of significance selected within the value range.

Usually the arithmetic mean or some other measure of central tendency is selected as the medial base.* Using equal classes within the series on each side of the medial base is the simplest approach, but unequal spans within each series may be warranted under certain circumstances to be considered later. Similarly, the number of classes on each side of the medial base may be equal or unequal, though equal is preferable when appropriate. The value key should graphically show the span per class.

As earlier suggested, the medial base may be any significant figure including zero if both negative and positive numbers are involved—as when mapping losses and gains over time for such subjects as population or production. Or it might be 100 percent if values both below and above 100 percent were involved. It might simply be a round, easily remembered number, such as the year 1900 if mapping the age of buildings in an urban neighborhood, if the number has genuine importance. A *range* might be used as a "medial class." However, as such a class would presumably be of a span different from the classes on either side, the resulting map might appear rather complicated—with at least three different unrelated class spans. The value key would have to be particularly well designed and annotated.

To illustrate this method of unequal classing, we will use the low value curve of Alternative 2B, with 20.6, the arithmetic mean of the 100 values, serving as the medial base. For six classes—three equal spans below and three equal spans above the mean—the classes would be as follows: 0–<6.9, 6.9–<13.7, 13.7–<20.6; and 20.6–<47.1, 47.1–<73.6, 73.6–100. Note that with the low value curve used, the class spans above the medial base are predictably much larger than those below.

If one desired to emphasize relationships to the mean, the values could

---

*Any value significant to the subject being mapped may be employed. When locations vary significantly in size or population, the *geographic mean* or *demographic mean*, reflecting the areas or populations of the locations, may be used. Sometimes the mid-value of the range will serve.

be converted to minus and plus quantities relative to it. In that event, the mean is set equal to zero and the class spans would be as follows: −20.6–<−13.7, −13.7–<−6.9, −6.9–<0; and 0–<26.5, 26.5–<53.0, 53.0–79.4. The title of such a map would have to be worded so as to make the relative nature of the values clear. For example, if the map portrayed per-acre wheat production in Kansas (by counties) relative to the mean for the state as a whole, the title might be "Kansas Wheat Production Above and Below the Per-Acre State Average, by Counties." A map of this kind would forcefully emphasize the difference between the more and less successful farming operations, and stress each level of success relative to the average. When this approach is used, the original value of the mean should be stated on the map somewhere. In contrast, if only positive values are used, the emphasis will be on the yields in absolute terms; the average yield might only be revealed in a notation regarding the value at the medial base.

A map prepared according to this negative/positive approach is shown in Figure 10–17, using the illustrative low value curve. It employs plane interpolated symbolism in two darkness levels, each divided into three equal classes defined by dashed contour lines similar to those used earlier in Figure 10–12.

If the values had been left all positive, the map would have remained the same except for the value key. However, the symbolism for the two lower value contour lines might have been reversed so that the shorter dashes applied to the smaller value of 6.9 and the longer dashes applied to the larger value of 13.7. The choice between these two procedures is a matter of personal preference: one method emphasizes the relationship to the mean; the other points out the relationship to the original values.

The medial base approach used with this combination type of symbolism may be employed for as many as ten classes (five on each side). With that number of classes, it could be used effectively with straight-line value curves as well, and classing could be equal throughout.

Instead of the combination method of symbolism illustrated, colors of varying hue might be used (cost permitting). Varying darknesses of red, for example, might be used for quantities below the base, with greens or grays above. Using black-and-white only makes it difficult to differentiate satisfactorily between the values above and below the medial base. Another reasonable solution would be to use darkness tones composed of variously shaped elements. For example, the tone elements might be jagged for values below the medial base, and rounded for those above. Such maps have been made in six classes by computer line printer with a fair degree of success.* Or darkness tones composed of lines might be used for values below the medial base and darkness tones of dots for those above. In general, however, linear darkness tones prove unsatisfactory and should be avoided insofar as possible.

Similar solutions might be used with conformant symbolism. However, the tones generally appear in diverse combinations with such symbolism

---

*The three jagged elements were +; ×; and < and > superimposed. The three rounded elements were 0; 0 and − superimposed; and 0, X, A, and V all superimposed.

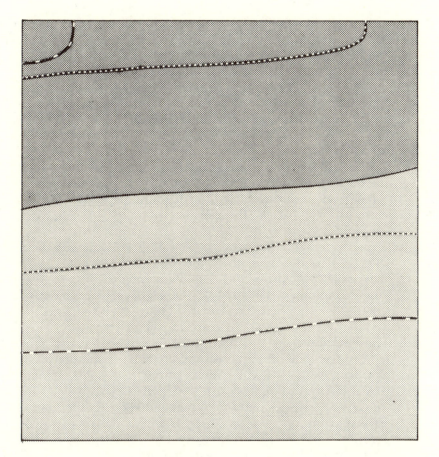

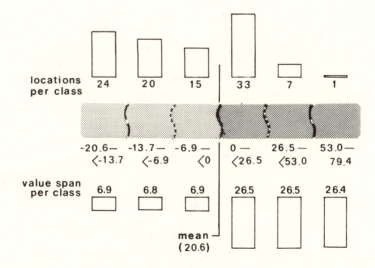

Figure 10–17   Classing Relative to the Mean, Low Value Curve (8-FD7)

rather than in quantitative progression as with interpolated symbolisms, and the map would be more difficult to comprehend. Even with varying hue, it might be difficult to use more than three classes on either side of the medial base.

The system of classing relative to a medial base may occasionally be used in combination with what is called *open-ended classing*, but in such instances ready comprehension tends to be difficult.

Open-Ended Classing

A rather special approach is sometimes used when well-rounded class limits are desired. For thematic maps intended for popular use by hurried newspaper readers, for instance, drastic rounding of class limits may at times be essential. In this case, the use of open-ended classing can prove extremely helpful.

In the previous discussion of rounding we dealt with the problem in terms of equal classing. When the value curves render equal classing unacceptable, it is quite difficult to find a solution capable of being expressed in well-rounded numbers. Unequal class breaks, well chosen when judged by other criteria, tend to yield ragged class limits. Well-rounded class limits tend to yield erratic class spans. If graphed, their curves would be characterized by kinks and reversals difficult for the map user to comprehend.

An ingenious, though not entirely happy, solution has been developed that ignores the extreme values entirely and sidesteps questions about the actual use of the highest and lowest classes. The first and last classes (or either) are simply shown as "open-ended," i.e., the lower limit of the first class and the upper limit of the last class are unspecified. In its simplest form, this method also permits the intermediate classes to be of equal span. This is possible because the extreme *stated* values can be chosen arbitrarily, so long as they fall within the range of the actual extremes.

The open-ended class concept may occasionally be used for other purposes than to permit well-rounded class breaks. For example, to map confidential information such as income levels, it might be necessary to suppress extreme values if using them could lead to the identification of the persons involved. Or extreme values of weather data might be de-emphasized when there is reason to believe that the lowest or highest temperature had been or might be exceeded at some time. When extremes are used as a basis for classing, they usually control the class spans, and any change in either or both extreme values would cause a change in the class spans. With open-ended classing, previously established class spans can continue to be used regardless of new facts, and the map designer gains great freedom of action. Instead of needing a classing curve that fits well over its entire length, a reasonable fit is needed only in part. Any serious troubles can all be hidden in those mysterious extreme classes. In a word, the extreme classes no longer need make sense.

Let us here examine the procedure for the assumed low value curve, using five classes. The values chosen for the *stated* extremes might be 5 and 35 (as compared to the actual extremes of 0.3–100). The resulting classes would then be as follows: <5, 5–<15, 15–<25, 25–<35, and >35. A map based upon such classing is shown in Figure 10–18. Since, for all the map user

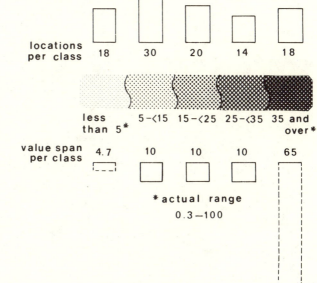

Figure 10–18 Open Ended Classing, Low Value Curve (8-FD7)

knows, the first class may start with any value more than negative infinity, while the last class may end with any value less than positive infinity, information as to the actual extreme values should be provided somewhere— as in, say, the footnote included in the value key.

If the user's interest were certain to be limited to a specific range of values, it might be appropriate for all locations with values beyond that range to be entirely removed from consideration, and the map title revised to reflect what is actually being presented. For example, if concern is with housing within a depressed area suitable for rehabilitation, information about parcels with housing requiring only normal repair and parcels with housing too far deteriorated to justify rehabilitation might be entirely omitted, and the map entitled "Housing Suitable for Rehabilitation, by Parcels."

With COUNT symbolism, including dot maps, the symbol is normally given a rating. The resulting map is ordinarily considered open-ended. For example, in Figure 9–8A, the span of the last class might be 98–102, while if the subject permitted negative values, the span of the first class might be $-2-<2$, or the overall value range could be less than 0–100.

While the concept of open-ended classing cannot be used with totally unclassed symbolism, it is possible to employ it through a combination of classed and unclassed symbolism. By way of example, assume the use of circle symbolism for a value set in which the minimum value is a small percentage of the maximum. For any given maximum circle size, the minimum values might require circles too small to produce with accuracy or to be read effectively. An open-ended class could be employed, symbolized by the smallest circle practical conforming in size to the upper limit of the class.

If the circle symbolism of Figure 9–4 were employed for the low value curve of Alternatives 2A and 2B, a value of 1.0 would probably be the smallest which could be sensibly represented. The four lowest values in the set are: 0.3, 0.5, 0.8, and 1.0. If unclassed circles were used for all values of more than 1.0, a classed circle (of a size suitable for 1.0) could be used for all smaller values. This procedure is regularly used, but should be mentioned in the value key or a related note.

Occasionally, a similar procedure might be justified for use with high values. With the previous value set, the three highest values are 62.5, 68.4, and 100.0. If unclassed circles were used for values of less than 60.0, a classed circle (of the largest size suitable) could be used for all larger values. This procedure would permit all the circles (except that representing 100.0) to be made significantly larger, and in consequence provide greater differentiability among most—without losing any differentiability whatever among the three largest (Figure 10–19).

With EXTENT symbolism it is normally best to use the value at the high end of any class span in establishing the size to represent that span. Here, however, the value at the low end is employed for the upper class.

Occasionally, it may be useful to employ classed symbolism for values that are relatively close together and unclassed for those relatively far apart. This might be done if circle or bar symbolism were desirable and the value curve was very low. For example, in mapping municipal population in a region characterized by many villages and towns and only a few large cities,

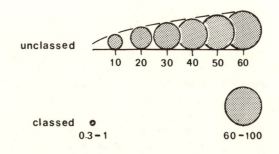

Figure 10–19 Value Key with Classed and Unclassed Circles

the latter might be shown unclassed with all others classed. Once again, careful thought must be given to the design of the value key.

Other Unequal Classing Procedures

There are other unequal classing systems, such as classing by specific mathematical curves and classing by value curves with major reversals. Perhaps the most useful unequal classing system, however, is that based on "descriptive categories," i.e., used with ordinal scaling. In the example used earlier, a map of population per square mile for a state by counties, it is possible to employ categories described by the words *rural, exurban, suburban,* and *urban,* rather than strictly numerical terms. Each word determines a class of increasing population density.

To prepare this map, the range of values considered characteristic of each descriptive category is first established. If any two overlap, a reasonable breakpoint can be chosen between them to avoid having any one location fall in more than one class. Likewise, if any two ranges fail to meet, the midpoint between them (or some other appropriate point in terms of rounding) can be established as the breakpoint to avoid an undefined gap between classes. Population per square mile should be stated for each class in the usual way, but in addition, each of the four classes should be assigned a particular descriptive identification to avoid confusion due to irregularity among the class spans.

This method is something like the quantile system, in that the classing will usually be unequal and without logical progression in the class spans. However, since there is a continuous quantitative flow from minimum to maximum values, interpolation can be appropriately employed, if desired— which is not usually the case with quantile classing because of its tie to locations rather than to quantities.

One more unequal classing "system," if it can be so called, exists. This is the mapmaker's intuitive judgment. Until relatively recently, this seems to have been the most commonly used procedure, and in consequence, most unequally classed maps have proven needlessly difficult to interpret. Few users will expend the time and effort required to understand a map based on what appear to be mysteriously shifting class intervals.

In the past, even under the best circumstances, graphic indication of the relative size of the class spans was seldom provided. Yet it is frequently difficult to comprehend classing from the numerical specifications of the class spans. This is especially true when the breakpoints between classes are expressed in ragged figures, and when the trend in class span is not continuously progressive. Confusion can be caused, for example, by something as simple as the use of two or more classes of identical span in what otherwise appears to be a smooth progression of increasing spans.

This difficulty also arises with quantile classing. If the equality which is present (usually the number of locations per class) is not stressed—either through suitable wording, or preferably, through a frequency distribution (as in the upper bars of Figure 10–14)—the map user is likely to have difficulty comprehending the classing.

Natural Breaks Another approach frequently suggested, though not frequently used, is based upon what are called "natural breaks." Most value curves have at least small irregularities, such as those illustrated in the upper chart of Figure 10–3. When such irregularities are substantial, they are sometimes used as a basis for establishing class breakpoints, on the theory that values of similar character (however defined) should be grouped together within the same class and that values of dissimilar character should *not* be grouped together. However, as this concept is diametrically opposed to the postulates supporting unequal classing, its validity is questionable. This is especially true since "natural breaks" have no existence except in the artificial situation that occurs when the values are arranged in order of magnitude. Even relatively minor locational departures from that nonspatial sequence seriously challenge the validity of the entire concept, especially if interpolation is to be employed.

With the natural break system of classing, the question that must always be asked is whether a so-called break in the value curve has any significance in terms of the value distribution *as it exists in space*. For example, imagine that terrain height is to be mapped by contouring. Would the presence of a large, slightly sloping stretch located between large, steep stretches in the value curve justify abandoning contour ratings that might otherwise be most appropriate under the circumstances? Even if the value curve corresponded precisely to a vertical profile through the terrain, i.e., if spatial sequence corresponded to ordered sequence, would it be desirable to show no contours within the areas of great or small slope—and to have contours (probably with odd value ratings) only at the points where slope changed?

While this example assumes interpolation and a physical surface, these questions should be asked under all possible circumstances. The erratic class spans that usually result from the use of natural breaks may be more conspicuous with plane interpolated symbolism, and hence less readily tolerated in those instances than with conformant symbolism. But they should be avoided in other cases as well. As was previously stressed, when numerous values fall close together—as at the low end of a typical low value curve—it is helpful to divide them among several classes. It is thus possible to distinguish among them and, at the same time, avoid having too large an area of the map undifferentiated. Employing the natural break concept, however, is likely to have the opposite effect, and its use is therefore best confined to special situations when its value and validity are not in doubt. These rough guidelines can be followed, though more research is needed on this subject.

An example of circumstances in which natural breaks may prove valid is a situation in which the class groupings can be descriptively categorized, as discussed above. The descriptive category approach may be used with any value curve, but sometimes the breakpoints between categories may coincide with natural breaks in the value curve. For instance, the value curve might show natural breaks at one or more of the points where urban, exurban, and suburban categories meet.

Sometimes one or more natural breaks may be used in combination with equal classes related to the breaks. For example, if reversal in a value curve were to be construed as a natural break, it might be used with equal but dissimilar classing on each side of it. (See Classing for Value Curves with Major Reversal, Appendix 3.) Similarly, if there were only one especially sharp bend in a value curve, it might be used in the same way. For example, if the value curve were quite low and at the same time strongly right-weighted, it might be desirable to use several equal classes for all values up to and including that at the point of maximum curvature, ignoring minor irregularities. One class of greater span could then contain all values greater than that at the point of maximum curvature. This could be made an open-ended class, if warranted, a procedure somewhat equivalent to using a dominant natural break as a medial base. Two especially sharp bends (with or without reversal) might under some circumstances be a basis for establishing open-ended classes with one set of equal classes between.

In any case, whenever natural breaks are employed, an effort should be made to categorize the resulting classes in some way that will be meaningful to the map user. If this cannot be done with words, the map should include a value curve diagram with the classing indicated by horizontal lines. Such a diagram can be extremely small and still reveal the rationale behind the classing.

Questions

I. Suppose you have 100 locations for which you wish to use quintile classing, and the first 25 locations all happen to have values of 0.
    1. What could you do to resolve the problem?
    2. How many locations would you place in each class?
II. What is wrong with these class spans, and why?
    1. 0–20, 20–40, 40–60, . . .
    2. 0–19, 20–39, 40–59, . . .
    3. −20–0, 0–<20, 20–<40, . . .
    4. 0–17.5, 21.9–38.2, 40.1–50.0, . . .
III. Is a contour map unclassed, classed, or both?
IV. High value curves can frequently be converted to low value curves.

        1. How can this be done?

        2. Why might you wish to do it?

V.   Circle symbolism, though usually used unclassed, is sometimes classed. For what reasons and under what circumstances might it be desirable or necessary to class circle symbolism? Why?

VI.   If the minimum value were a large percentage of the maximum value and you wished to use bar symbolism, what would you do differently than if the minimum value were a small percentage of the maximum value? Why?

VII.   Assume the use of five classes for a map of 100 locations with a value range of 0–100, 70 of which fall within 4 value points of one another. Would it in general be preferable:

        1. To represent all 70 locations by the same symbolism—on the grounds that they are close together in value and therefore should be symbolized the same, or

        2. To use two or more different symbolisms—on the grounds that there are so many values close together they should be differentiated?

VIII.   In mapping the surface of the earth, pits are infrequently encountered because they tend to fill up and become lakes. In thematic mapping generally, are pits and peaks equally likely to be encountered? Explain.

IX.   With classed symbolism, what objection, if any, is there to having an empty class (i.e., a class in which no location falls)? Why?

X.   Under what circumstances might it be reasonable to class with 40 classes?

XI.   Extreme classes are seldom empty, though intermediate classes may be. Under what circumstances might one or both extreme classes be empty?

XII.   Assume that you are using a census tape to make a series of computer maps of a study space with a view to showing a variety of different subjects related to city planning. If a map came out correctly all in one class, would you consider it worthless? If not, why not?

Answers

I.   1. Adjust map title to exclude locations of 0 value.

        2. 15.

II.   1. Where would values of exactly 20, 40, etc. fall? They could go in either of two classes.

        2. Where would values of 19.5, 39.2, etc. fall? They are not included in any class.

        3. Where would a value of 0 fall? It can be placed in two classes.

        4. They are difficult for the user to easily understand, and certain values within the entire range would fall between classes.

III.   Both. It is classed because there is no differentiation by value with any tone. It is unclassed in that the position of each tone is determined by unclassed values. Tones meet at the equivalent of a contour line and countour lines as symbolism representing a specific value are unclassed.

IV. 1. Convert high value curves by adjusting the subject so that the curve becomes low (i.e., the subject of illiteracy rather than literacy in a developed area).

2. A map with many low values and few high values is easier to make and interpret. It is easier with circles and bars to differentiate among a number of small symbols than among a number of large symbols. The map is less cluttered, and with darkness maps information can be more effectively presented.

V. Classing is desirable when the minimum value is a fairly large percentage of the maximum value, to provide better differentiability among the circles. It is also desirable when many locations are involved, to save time in drawing circles of varying size.

Classing is necessary when the minimum value is a very large percentage of the maximum, to provide differentiability among the circles. Partial classing is necessary if the smallest circle that can be drawn or serviceably used is too large for the proper representation of some values.

VI. Use equal classing based on the class numbers, to provide better differentiability among the bars.

VII. It would be preferable to use different symbolisms, because differentiability among numerous values close together is usually desirable.

VIII. With a straight value curve, pits and peaks are probably equally likely to be encountered. However, most value curves are low, so overall there is probably a likelihood of there being more pits than peaks, since with more low values there will be a greater tendency for them to merge.

IX. Usually it is desirable to have approximately the same number of locations in each class. An empty class is inconsistent with that objective. Otherwise, there is probably no objection to an empty class with interpolation. With plane conformant symbolism, it is best to have every class represented frequently on the map, as the tones can then be better judged. For example, if five classes are used, the absence of one tone may create some uncertainty as to which four tones are present.

X. You might wish to use 40 equal classes if the value curve was very low and you wished to avoid unequal classing. Forty classes could be represented by the use of COUNT or DOT symbolism.

If you wished to show data very accurately yet generalize to some extent, you could use SPOT-COUNT symbolism. This symbolism provides overall comprehension at a glance, yet also enables the user to determine the approximate value applicable to any particular location with ease by counting. The level of accuracy is quite high since class spans are small. This type of symbolism would prove useful if you wanted to map something such as total population.

XI.   Extreme classes might be empty if the same class spans are used for many (ten or more) maps with different value ranges.

Extreme classes might also be empty in a situation in which it is desirable to show that a possibility exists for unusual data. For example, in classing I.Q. one may wish to leave a part of the last class empty when the odds are good that someone with a very high I.Q. may fit in that class, although there may be no such person in the particular sample given.

XII.   No. It would show for the subject represented that all of the values were identical (or that they were in the same class, if specific class spans had been preestablished).

CHAPTER 11

Single-Subject Mapping

PROCEDURE

Editor's note: The questions and answers below form a procedural outline for single-subject mapping. The numbered footnotes will be referred to repeatedly.

Assuming the values as given are according to NOMINAL SCALING:

I. Is the study space to be differentiated (other than by value symbolism)?[1]
   A. If so, assign it to its base area for symbolization.
   B. If not, ignore.

II. Are there any given locations to which the given nominal categorization does not apply?
   A. If so, and such locations are to be differentiated from space not within locations, assign all locations. (See III below.)
   B. If not, assign only locations of presence.

III. What type of location assignment is to be used?
   A. If assignment is to base areas:
      1. For presence, use FIELD symbolism.
         a. If of DARKNESS type, apply a distinctive tone over each base area (or inside around the perimeter).[2]
         b. If of EXTENT type, raise the base area of each location.[2]
      2. For absence, use FIELD symbolism.

[1]Differentiation (if required) is usually accomplished by the use of an outline around the study space, or by the use of a contrasting tone within or beyond the study space. If the study space is not to be differentiated, all locations should usually be differentiated.

[2]If individual locations are to be differentiated, outline each base area as necessary. (When assignment is to centerlines, outline them.)

      a. Use white paper.[2] [3]

      b. Or, with DARKNESS type, use a light tone contrasting well to that used for presence.[2]

  B. If assignment is to base centerlines:

    1. For presence, use BAND symbolism.

      a. If of DARKNESS type, apply a band of distinctive tone centered on each base centerline.[2]

      b. If of EXTENT type, raise the base centerline (to form a "fin").

    2. For absence, use BAND symbolism.

      a. Use white paper.[2]

      b. Or with DARKNESS type, use a band of light tone, contrasting well to that used for presence, centered on each base centerline.[2]

  C. If assignment is to base centers:

    1. For presence, use SPOT symbolism.[4]

      a. If of DARKNESS type, apply a distinctive tone over each symbol area.

    2. For absence:

      a. Use white paper.[5]

      b. Or use a light tone, contrasting well to that used for presence over each symbol.

      c. Or use a small dot.

Assuming the values as given are according to ORDINAL SCALING:

I. Is the study space to be differentiated (by other than value symbolism)?

  A. If so, assign to its base area for symbolization.[1]

  B. If not, ignore.

II. What type of location assignment is to be used?

  A. If assignment is to base areas, use FIELD symbolism.

    1. If of DARKNESS type, apply tones of increasing darkness[6] over each base area or inside along its perimeter.

    2. If of EXTENT type, increasingly raise each base area.[2]

  B. If assignment is to base centerlines, use BAND symbolism.

    1. If of DARKNESS type, apply bands of tone of increasing darkness[3] centered on each base centerline.

---

[3]If overall areas of absence are to be indicated (without the base areas being individually differentiated), outline them as necessary.

[4]Other than of volumetric type, centered on (or otherwise consistently related to) the base centers.

[5]If individual locations are to be differentiated, outline a corresponding symbol shape at each base center.

[6]For each category of increasing significance, proceed by well-differentiated steps according to the concept of ordinal ranking. There will usually be a choice between ranking in the direction of increasingly desirable conditions and ranking in the direction of increasingly undesirable conditions.

2. If of EXTENT type, apply bands of tone of increasing width or height centered on each base centerline.

3. If of COUNT type, use an increasing number of lines within each band.

C. If assignment is to base centers, use SPOT symbolism.[3]

1. If of DARKNESS type, apply tones of increasing darkness over each symbol area.

2. If of EXTENT type, increasingly enlarge each symbol.

3. If of COUNT type, use an increasing number of elements within each spot.

Assuming the values as given are according to INTERVAL or RATIO SCALING:

I. Is the study space to be differentiated? (If the values are to be mathematically adjusted or classed, see III and IV below.)

A. If so, assign to its base area for symbolization.

B. If not, ignore.

II. What type of location assignment is to be used?

A. If assignment is to base areas, use FIELD-type symbolism.

1. If of DARKNESS type, apply tones of increasing darkness[2] (usually requires classing; see IV below).

2. If of EXTENT type, use increasing height.[2]

3. If of COUNT type, use an increasing number of symbol elements (requires classing; see IV below).

B. If assignment is to base centerlines, use BAND-type symbolism.

1. If of DARKNESS type, use increasing darkness[2] (usually requires classing).

2. If of EXTENT type, use increasing size.[2]

3. If of COUNT type, use an increasing number of symbol elements (requires classing; see IV below). If interpolation is to be used, assign locations also to base centerlines and perform linear interpolation along base centerlines.

4. If of DARKNESS type, use gradually increasing darkness within bands.

5. If of EXTENT type, use gradually increasing size of bands.

6. If of COUNT type, use an increasing number of symbol elements (requires classing; see IV below).

C. If assignment is to base centers, use spot type symbolism (unless interpolation is to be employed).

1. If of DARKNESS type, use increasing darkness[2] (usually requires classing; see IV below).

2. If of EXTENT type, use increasing size.[2]

3. If of COUNT type, use an increasing number of symbol elements (requires classing; see IV below). If interpolation is to be used, perform areal interpolation over study space.

4. If symbolism is of DARKNESS type, use increasing darkness (usually requires classing; see IV below).[3]

5. If symbolism is of EXTENT type, use increasing height.[2]

6. If symbolism is of COUNT type, use an increasing number of symbol elements (requires classing; see IV below).

III. Are the values to be mathematically adjusted?
  A. If so, how?
     1. By roots.
     2. By powers.
     3. By log.
     4. By other means.

IV. Are the values to be classed?
  A. If so, are the classes to be equal or unequal?
     1. If unequal, how?
        a. Quantile.
           i. By location.
           ii. By areas.
           iii. By other.
        b. Reciprocal curve classing.
        c. By specific mathematical curves (based on value extremes).
           i. Arithmetic.
           ii. Geometric (logarithmic).
           iii. Reciprocal.
           iv. Other.
        d. Relative to a medial base.
        e. Descriptive categories.
        f. "Natural breaks."
        g. Other.
  B. How many classes are to be used?
     1. If few (2–5), all symbolisms and quantitative analogues.
     2. If many (6–10), all symbolisms and quantitative analogues except bars (as SPOT-COUNT).
     3. If very many (more than 10), all symbolisms and quantitative analogues except bars or sectors (as SPOT-COUNT).

If zero is one of the values and it is to be treated as the equivalent of absence, see under NOMINAL SCALING for alternative procedures for its treatment.

For each value, unless classing is employed (see IV), the quantitative analogue of the value symbolism should in theory conform to the value to be represented. This presents no problem when EXTENT is the analogue. However, when DARKNESS is the analogue, the range of darkness is usually made to conform to the range of the values, regardless of the actual magnitude of the values.

Sparse France: A Problem with Alternative Solutions

Having considered symbolism in Chapter 9, let us now consider its application to a real-life problem with an actual study space, governmentally-established locations, and national census values applicable to the locations.

The Study Space, Locations, and Values

Suppose we are interested in mapping France, that portion of the coterminous nation which has a population density of less than 500 persons per square mile. We will refer to this as Sparse France. Our interest is in the number of voters that were registered in each department on February 28, 1970.

For assignment, since we will be producing small maps, the outlines of the areas must be substantially simplified. We cannot attempt to represent the outer limits of every rock, pebble, and grain of sand defining the coastline of France—or even many far grosser features such as harbors and smaller peninsulas. (A considerable degree of generalization is frequently a virtue to be deliberately sought rather than a necessity to be reluctantly tolerated.)

Turning now to the locations, we find that 82 out of the 94 departments of coterminous France had a population in 1970 of less than 500 persons per square mile. Figure 11–1 shows the study space and its 82 locations assigned to their base areas, with the locations identified by their reference numbers.

On this map, the 12 areas shown in gray represent those departments which had a population in 1970 of 500 or more persons per square mile, and are therefore not a part of the Sparse France study space. Here and elsewhere it has not been necessary to show a north arrow since north is up on the map. A graphic scale of distance has also been omitted for simplicity, as it is not essential to our purposes. The directional trend of adjacent coastlines has been suggested by short lines.

The location base areas of Sparse France vary to some degree in size, as is evident in Figure 11–1. The hollow bars in Figure 11–2 show area in square kilometers. The small gray bars show the percentage of the area in each case which is below the mean for all of the locations. The small black bars show the corresponding percentage which is above the mean. Where no small bars are shown, the area is within 10 percent of the mean. The largest gray bar thus indicates the smallest department (No. 24); the largest black bar indicates the largest department (No. 71). The combined symbolism for any location shows the relationship to the mean.

As stated in the value key, Haut Rhin, toward the upper right, has the smallest area (43 percent less than the mean), while Gironde, toward the middle left, has the largest area (62 percent over the mean). While the size of the latter is thus about 3.7 times that of the former, the differences existing here are very small compared to what is usually encountered in mapping work. As differences in location size increase, problems of map symbolization and interpretation also tend to increase.

With Foursquare, reference numbering presented no problem because of the regular layout of locations. With irregularly disposed locations (more commonly encountered), however, there is no ideal solution. The best procedure uses a back-and-forth sequence, across the study space from top to bottom. When reference numbers are established according to any other procedure, such as perhaps in accordance with the alphabetical sequence of the departmental names, the task of finding a location corresponding to any particular reference number can become quite difficult.

Figure 11–1   France with Locations Assigned to Base Areas

The Value Set and the Value Curve

Table 11–1 provides the information to be portrayed for each location, listed in order of increasing magnitude. This is usually the most useful order to employ if only one listing is to be made. From such a listing the applicable value curve can most easily be constructed, and quantile classing can be established if desired. One alternative is a listing according to reference number sequence. This ordering is helpful when working from a map marked with reference numbers, such as Figure 11–1, and is commonly used with computer-generated maps. Another alternative is to list alphabetically the names of the zones. Although least important, this ordering can prove useful when the locations have established designations and their number is relatively large.

The subject for mapping is "Registered Voters, by Departments—1970." Since the values are of addable type, each location will have to be differentiated on the map. However, with several of the symbolisms we wish to

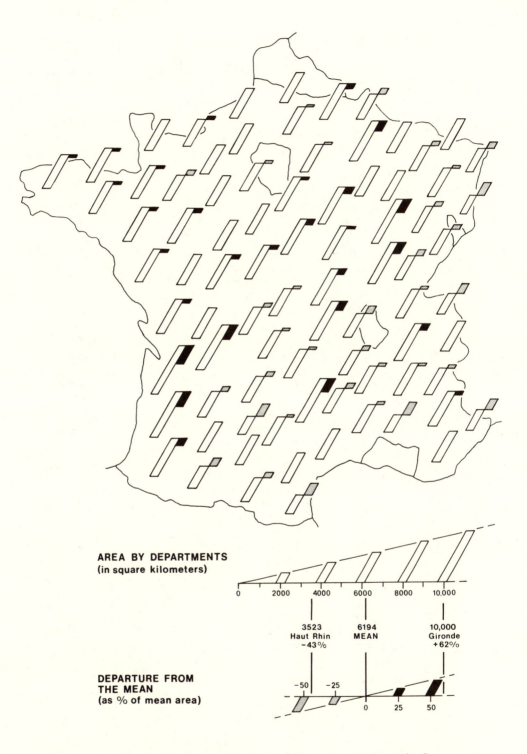

Figure 11–2   Symbolic Representation of Department Areas in Sparse France

Table 11–1: Sparse France: Registered Voters, by Departments—1970

Increasing Value Sequence

MAP REFERENCE NOS.	LOCATIONS	VALUES	
67	Lozère	52,630	
61	Hautes Alpes	59,252	
62	Alpes-de-Haute-Provence	66,102	
79	Ariège	94,813	
69	Lot	101,956	
57	Cantal	110,484	
48	Creuse	110,546	
74	Tarn-et-Garonne	113,700	
73	Gers	115,314	
5	Meuse	121,693	
10	Haute Marne	122,751	
23	Haute Saône	132,997	
58	Haute Loire	137,406	
81	Hautes Pyrénées	142,043	
26	Jura	145,036	
32	Mayenne	155,626	
28	Nièvre	156,746	
11	Aube	156,794	
40	Indre	162,220	
56	Corrèze	162,400	
4	Ardennes	166,277	
30	Loire-et-Cher	166,810	
59	Ardèche	167,377	
44	Savoie	167,400	
18	Orne	171,872	
78	Pyrénées-Orientales	172,014	
21	Yonne	176,660	
77	Aude	177,165	
70	Lot-et-Garonne	179,068	
19	Eure-et-Loir	181,595	
72	Landes	181,949	
68	Aveyron	188,174	
29	Cher	189,092	
51	Deux Sèvres	202,503	
42	Ain	203,921	
60	Drôme	205,134	
65	Vaucluse	206,440	
54	Charente	206,730	
50	Vienne	209,192	
75	Tarn	211,813	
43	Haute Savoie	216,983	
15	Eure	224,130	median:
49	Haute Vienne	229,605	229,557
25	Doubs	230,133	
9	Vosges	233,955	
22	Côte-d'Or	239,674	

Table 11–1: Sparse France: Registered Voters, by Departments—1970 *(continued)*

Increasing Value Sequence

MAP REFERENCE NOS.	LOCATIONS	VALUES	
41	Allier	243,702	mean:
55	Dordogne	253,424	249,161
20	Loiret	255,496	
39	Indre-et-Loire	255,665	
52	Vendée	263,220	
12	Marne	270,150	
17	Manche	271,360	
31	Sarthe	275,228	
66	Gard	288,510	
16	Calvados	296,508	
3	Aisne	296,978	
53	Charente-Maritime	298,138	
14	Oise	298,322	
2	Somme	304,455	
34	Côtes-du-Nord	314,440	
82	Pyrénées-Atlantiques	318,115	
47	Puy-de-Dôme	324,639	
27	Saône-et-Loire	334,994	
64	Var	335,288	
36	Morbihan	340,566	
38	Maine-et-Loire	342,404	
76	Herault	342,615	
24	Haut Rhin	345,170	
13	Seine-et-Marne	348,184	
6	Meurthe-et-Moselle	380,962	
33	Ille-et-Vilaine	396,852	
80	Haute Garonne	402,225	
46	Loire	417,998	
45	Isère	424,589	
63	Alpes-Maritimes	433,734	
8	Bas Rhin	472,740	
7	Moselle	500,678	
35	Finistère	502,828	
37	Loire-Atlantique	510,750	
71	Gironde	602,173	
1	Seine-Maritime	641,917	
	TOTAL	20,431,192	

Reference Number Sequence

MAP REFERENCE NOS.	LOCATIONS	VALUES	
1	Seine-Maritime	641,917	max.
2	Somme	304,455	
3	Aisne	296,978	
4	Ardennes	166,277	
5	Meuse	121,693	
6	Meurthe-et-Moselle	380,962	

Table 11–1: Sparse France: Registered Voters, by Departments—1970 *(continued)*

Reference Number Sequence		
MAP REFERENCE NOS.	LOCATIONS	VALUES
7	Moselle	500,678
8	Bas Rhin	472,740
9	Vosges	233,955
10	Haute Marne	122,751
11	Aube	156,794
12	Marne	270,150
13	Seine-et-Marne	348,184
14	Oise	298,322
15	Eure	224,130
16	Calvados	296,508
17	Manche	271,360
18	Orne	171,872
19	Eure-et-Loir	181,595
20	Loiret	255,496
21	Yonne	176,660
22	Côte-d'Or	239,674
23	Haute Saône	132,997
24	Haut Rhin	345,170
25	Doubs	230,133
26	Jura	145,036
27	Saône-et-Loire	334,994
28	Nièvre	156,746
29	Cher	189,092
30	Loire-et-Cher	166,810
31	Sarthe	275,228
32	Mayenne	155,626
33	Ille-et-Vilaine	396,852
34	Côtes-du-Nord	314,440
35	Finistère	502,828
36	Morbihan	340,566
37	Loire-Atlantique	510,750
38	Maine-et-Loire	342,404
39	Indre-et-Loire	255,665
40	Indre	162,220
41	Allier	243,702
42	Ain	203,921
43	Haute Savoie	216,983
44	Savoie	167,400
45	Isère	424,589
46	Loire	417,998
47	Puy-de-Dôme	324,639
48	Creuse	110,546
49	Haute Vienne	229,605
50	Vienne	209,192
51	Deux Sèvres	202,503
52	Vendée	263,220
53	Charente-Maritime	298,138
54	Charente	206,730

Table 11–1: Sparse France: Registered Voters, by Departments—1970 *(continued)*

Reference Number Sequence

MAP REFERENCE NOS.	LOCATIONS	VALUES	
55	Dordogne	253,424	
56	Corrèze	162,400	
57	Cantal	110,484	
58	Haute Loire	137,406	
59	Ardèche	167,377	
60	Drôme	205,134	
61	Hautes Alpes	59,252	
62	Alpes-de-Haute-Provence	66,102	
63	Alpes-Maritimes	433,734	
64	Var	335,288	
65	Vaucluse	206,440	
66	Gard	288,510	
67	Lozère	52,630	min.
68	Aveyron	188,174	
69	Lot	101,956	
70	Lot-et-Garonne	179,068	
71	Gironde	602,173	
72	Landes	181,949	
73	Gers	115,314	
74	Tarn-et-Garonne	113,700	
75	Tarn	211,813	
76	Herault	342,615	
77	Aude	177,165	
78	Pyrénées-Orientales	172,014	
79	Ariège	94,813	
80	Haute Garonne	402,225	
81	Hautes Pyrénées	142,043	
82	Pyrénées-Atlantiques	318,115	
	TOTAL	20,431,192	

Alphabetical Sequence

MAP REFERENCE NOS.	LOCATIONS	VALUES
42	Ain	203,921
3	Aisne	296,978
41	Allier	243,702
62	Alpes-de-Haute-Provence	66,102
63	Alpes-Maritimes	433,734
59	Ardèche	167,377
4	Ardennes	166,277
79	Ariège	94,813
11	Aube	156,794
77	Aude	177,165
68	Aveyron	188,174
8	Bas Rhin	472,740
16	Calvados	296,508

Table 11–1: Sparse France: Registered Voters, by Departments—1970 (continued)

Alphabetical Sequence

MAP REFERENCE NOS.	LOCATIONS	VALUES	
57	Cantal	110,484	
54	Charente	206,730	
53	Charente-Maritime	298,138	
29	Cher	189,092	
56	Corrèze	162,400	
22	Côte-d'Or	239,674	
34	Côtes-du-Nord	314,440	
48	Creuse	110,546	
51	Deux Sèvres	202,503	
55	Dordogne	253,424	
25	Doubs	230,133	
60	Drôme	205,134	
15	Eure	224,130	
19	Eure-et-Loir	181,595	
35	Finistère	502,828	
66	Gard	288,510	
73	Gers	115,314	
71	Gironde	602,173	
24	Haut Rhin	345,170	
80	Haute Garonne	402,225	
58	Haute Loire	137,406	
10	Haute Marne	122,751	
23	Haute Saône	132,997	
43	Haute Savoie	216,983	
49	Haute Vienne	229,605	
61	Hautes Alpes	59,252	
81	Hautes Pyrénées	142,043	
76	Herault	342,615	
33	Ille-et-Vilaine	396,852	
40	Indre	162,220	
39	Indre-et-Loire	255,665	
45	Isère	424,589	
26	Jura	145,036	
72	Landes	181,949	
46	Loire	417,998	
37	Loire-Atlantique	510,750	
30	Loire-et-Cher	166,810	
20	Loiret	255,496	
69	Lot	101,956	
70	Lot-et-Garonne	179,068	
67	Lozère	52,630	min.
38	Maine-et-Loire	342,404	
17	Manche	271,360	
12	Marne	270,150	
32	Mayenne	155,626	
6	Meurthe-et-Moselle	380,962	
5	Meuse	121,693	
36	Morbihan	340,566	

Table 11–1: Sparse France: Registered Voters, by Departments—1970 *(continued)*

MAP REFERENCE NOS.	LOCATIONS	VALUES	
	Alphabetical Sequence		
7	Moselle	500,678	
28	Nièvre	156,746	
14	Oise	298,322	
18	Orne	171,872	
47	Puy-de-Dôme	324,639	
82	Pyrénées-Atlantiques	318,115	
78	Pyrénées-Orientales	172,014	
27	Saône-et-Loire	334,994	
31	Sarthe	275,228	
44	Savoie	167,400	
13	Seine-et-Marne	348,184	
1	Seine-Maritime	641,917	max.
2	Somme	304,455	
75	Tarn	211,813	
74	Tarn-et-Garonne	113,700	
64	Var	335,288	
65	Vaucluse	206,440	
52	Vendee	263,220	
50	Vienne	209,192	
9	Vosges	233,955	
21	Yonne	176,660	
	TOTAL	20,431,192	

Source: Institut National de la Statistique et des Etudes Economiques, Annuaire Statistique de la France 1972, 77e Volume, Resultats de 1970, Nouvelle Serie No. 19, Ministère de l'Economie et des Finances, Republique Française, p. 125.

show, this cannot be done reasonably. Fortunately, the values can be converted to nonaddable type by the simple expedient of dividing each by the numeral 1 (one department), with adjustment of the title to "Registered Voters per Department, by Departments—1970." With this alteration, locational differentiation is no longer an essential requirement. (See Chapter 7, section on Addable and Nonaddable Values.)

Figure 11–3 shows the value curve for registered voters in Sparse France in 1970. It is somewhat low in comparison to that of Foursquare (Figure 10–3). The small reversal appearing at its lower left beginning is characteristic of a "lognormal" type of distribution. The slight irregularities in this and the Foursquare curve are, however, characteristic of most data based upon observations.

Along the left side of the chart the value range has been expressed not only in absolute terms but also by percentage between the extreme values. Showing such percentage figures is frequently helpful—and necessary whenever the reciprocal method of unequal classing is to be used. The mean and median quantities applicable to the 82 values have also been shown; sighting (or drawing) their levels across the chart will indicate their relationship to the values.

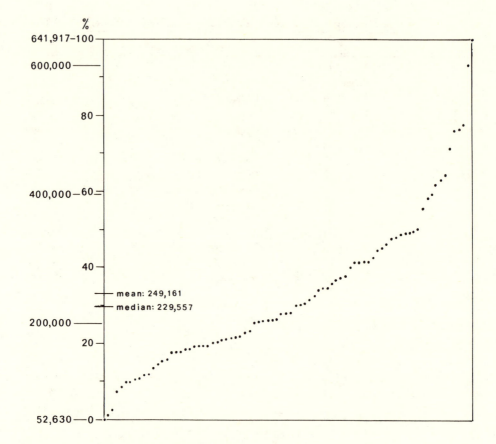

Figure 11–3   The Sparse France Value Curve

Symbolisms

Circles   Figures 11–4A, B and C use unclassed circles, with the area of each circle made proportional to the quantity being represented. Tone has been omitted from the dense areas to avoid confusion with the tones used to emphasize the circles. Also omitted is a graphic scale of distance.

These three maps use circles of three different size ranges. The maps here presented correspond to that for Foursquare shown in Figure 9–4. The base outlines of the locations are not shown. However, if considered sufficiently useful for any reason, they might be added, perhaps as thin light gray lines, so as to avoid overburdening the map. If added, the "spatial density" rate— registered voters per unit of area—might to some extent be indicated.

As to the best range of circle sizes to employ, there are no set rules and— as with most symbolisms of EXTENT type—the judgment of the map de-signer must determine the solution. In (A) the circles appear rather small and the differences among them difficult to judge. In (C) they appear to be too large and a rather crowded effect is produced. Location base outlines would crowd it even more. Under most circumstances the middle course shown in (B) would be desirable.

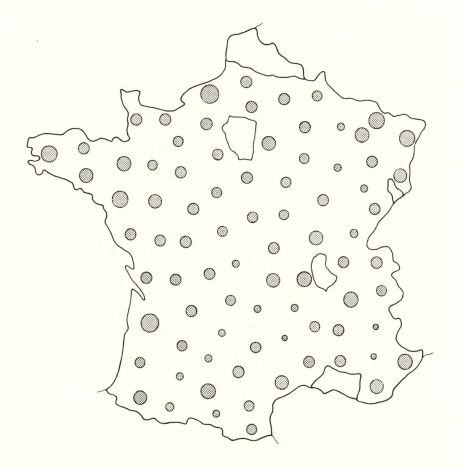

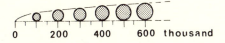

0   200   400   600   thousand

Figure 11–4A   Spot-Extent (1-SE3), Small Circle Size

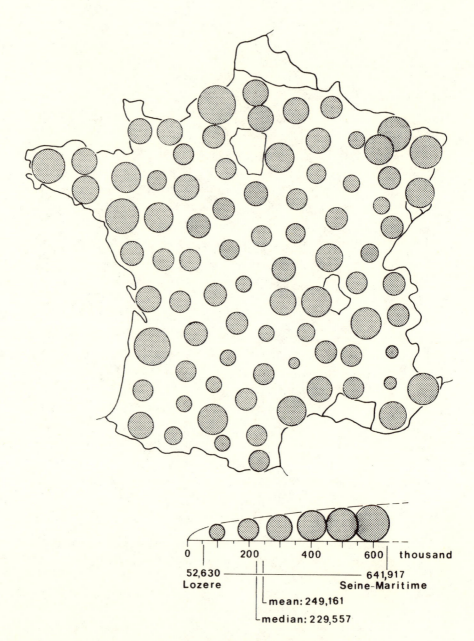

Figure 11–4B   Spot-Extent (1-SE3), Medium Circle Size

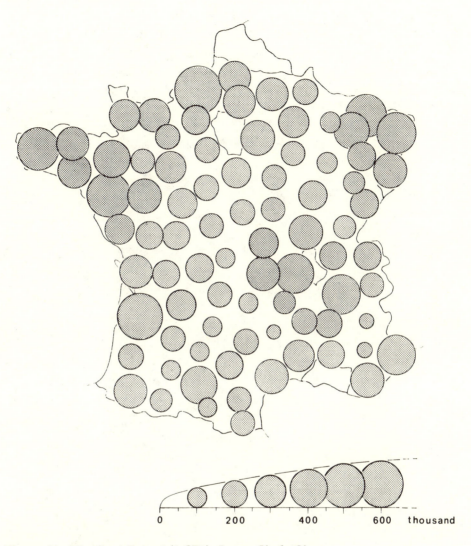

Figure 11–4C   Spot-Extent (1-SE3), Large Circle Size

Additional information of possible interest to the map user has been provided in the value key of the center map. The circumstances under which
such additional information might be included depend upon the judgment
of the designer or the wishes of the map sponsor. If the several kinds of
information shown here can easily be provided, it is desirable to include
them.

Bars   The three different maps in Figures 11–5A, B and C illustrate the effects
of varying bar length. Again the symbols in (A) appear rather small, while
(C) looks rather crowded.

While symbol overlap is less likely with bars than with circles, it is far

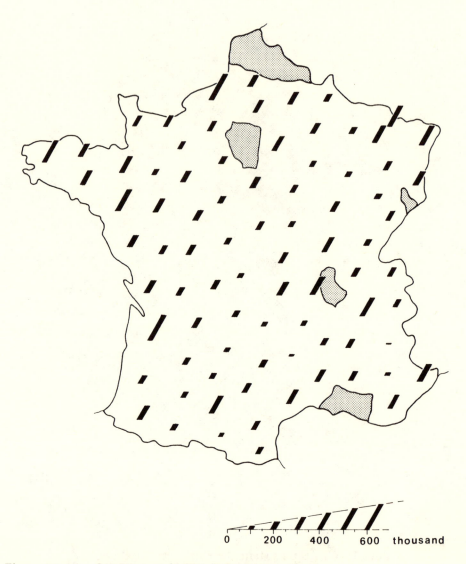

Figure 11–5A   Spot-Extent (1-SE1a), Small Bar Size

more serious when it occurs. Note in (C) at the middle left the conflict between the bars for locations 71 and 54. The angle at which the bars are drawn may sometimes be adjusted to minimize the problem. However, when bars are relatively long, conflicts may be difficult or impossible to avoid. In a situation of this kind, slightly repositioning a symbol may be justified.

Base location outlines might have been included here, in which case the shorter bars of (A) would probably be desirable. Bars of 3-dimensional form might have been employed here, drawn either by hand or by computer. The use of bars of cylindrical shape (perhaps foreshortened into an elliptical form) can be especially effective if the result is not too crowded.

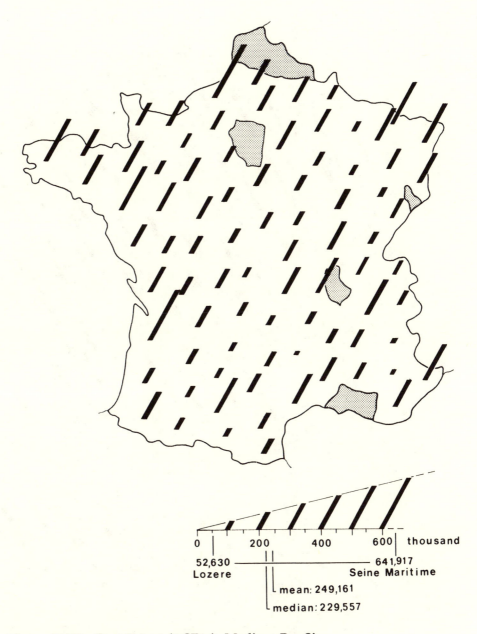

Figure 11-5B   Spot-Extent (1-SE1a), Medium Bar Size

Plane Conformant   We come next to FIELD symbolism of plane conformant type (Figure 11–6). Those areas that are not a part of Sparse France have again been left white, to avoid confusion with the tones representing values. With addable quantities it is essential that each separate location be differentiated, for which reason the departmental base outlines have here been shown in full. If the map subject had been registered voters per square

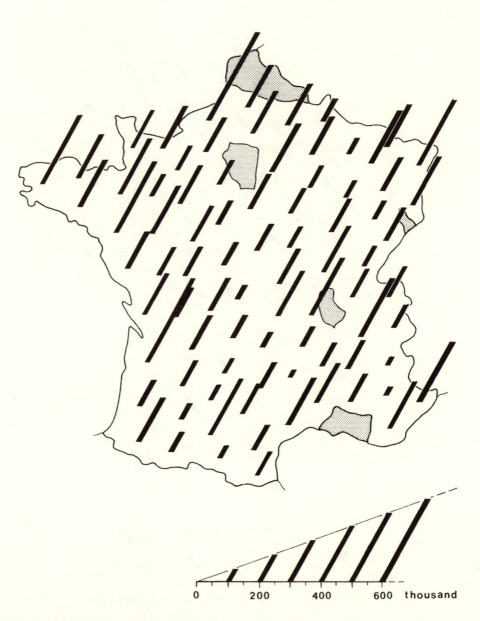

Figure 11–5C   Spot-Extent (1-SE1a), Large Bar Size

kilometer or any other unit of area, showing the base outlines would have been optional.

When the number of registered voters is being represented by locations, the darkness of the symbolism for each separate location (irrespective of its size) serves as the basis for interpretation. When the locations do not vary greatly in size, as in the present example, no difficulty is likely to be encountered. However, as we shall see (with Dense France), when locations vary significantly in size a serious problem may arise when this type of

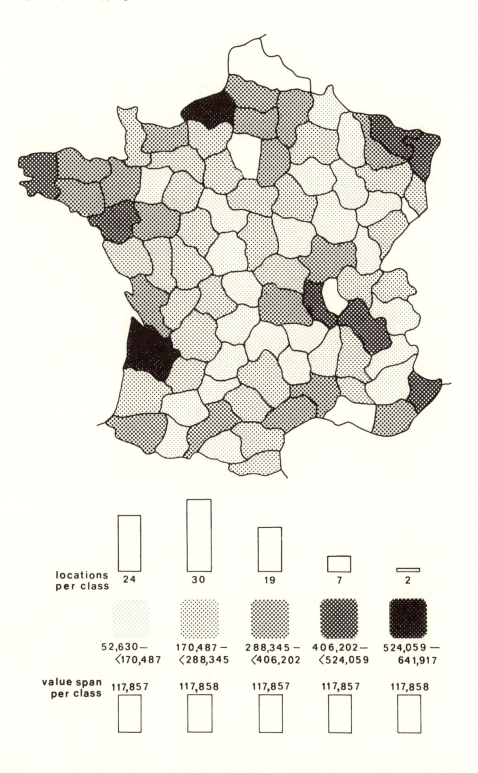

Figure 11–6   Field-Darkness: Plane Conformant (8-FD4), Five Equal Classes

symbolism is used with addable values. A given tone spread over a small location inevitably attracts less attention and tends to suggest the presence of a smaller quantity than the same tone spread over a large location. No such problem arises when the values are of nonaddable type.

In Figure 11–6, five classes of essentially equal span have been used, represented by tones of increasing darkness selected for maximum differentiability. With this type of symbolism the quantitative analogue of darkness is usually only roughly proportional to the quantity represented by each class.

The value keys employed may often appear too large for the size of the maps. In general, the larger the map the more appropriate it may be to include such supplementary information. Under some circumstances, the percentage of locations falling in each class (29.3, 36.6, 23.3, 8.5, and 2.4 in the present example) might also be included—and perhaps even the actual areas (and their percentages) falling in each class.

Figure 11–7 presents an alternative value key for the preceding map. To permit more rounded class intervals the extreme values have been slightly extended—from 52,630 down to 52,000 for the minimum value, and from 641,917 up to 642,000 for the maximum value. In consequence of this slight adjustment, five equal classes with spans of 118,000 each are possible without any change in the map. If the extremes had been further extended to achieve a range of 50,000–650,000, each class span would be the well-rounded figure of 120,000. In that event, however, adjustment would be required in the map, since department No. 66 (Gard), with a value of 288,510, would shift from Class 3 down to Class 2.

While the value curve applicable to registered voters in Sparse France (shown in Figure 11–3) is not especially low, it is low enough to produce, with equal classing, a major difference in the number of locations per class.

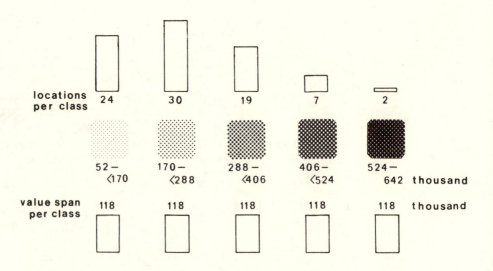

Figure 11–7   Value Key with More Extensive Rounding

This is shown by the great disparity in size among the upper bars of the value keys of the last two figures. As may be seen, the fifth class contains only two locations, in contrast to the second class which is fifteen times larger.

The quantile system of classing might be considered here. This is illustrated in Figure 11–8, in terms of five slightly rounded classes. Since the total number of locations is not evenly divisible by five (or by any alternative number of classes that might reasonably be considered), the number of locations per class necessarily varies by one. In the present example, the precise sequence was chosen by determining which configuration yielded the best rounding of the class span limits and at the same time was most nearly balanced.

Table 11–2 shows the range of possibilities. As may be seen, arrangement 8 (with 17 locations in the second and third classes) best meets the recommended criteria.

In the present example there are ten possible ways to place three classes containing 16 locations and two classes containing 17 locations. Each is shown in the table. The third column presents the unrounded class spans while the last column presents the rounded class spans. For the former, the breakpoints have been established at the midpoint between the highest value in the adjacent lower class and the lowest value in the adjacent higher class, with fractional results raised to the next highest integer. For the rounded class spans the breakpoints have been so established as to yield what appeared to be the best rounding. When there was a choice, that value nearest the midpoint was used.

Two of the arrangements (Nos. 1 and 8) permitted rounding to multiples of 1000—with four arrangements each to multiples of 500 and 250. Of the two that gave the best rounding, No. 8 provided a more even disposition. No. 6 gave the most balanced disposition but it provided less felicitous rounding. If rounding were not be employed, this arrangement would have been preferred.

Figure 11–9 shows simplified classing charts for equal and quantile classing.

Raised Conformant   Figures 11–10A, B and C provide an illustration of the use of unclassed raised conformant symbolism in three heights. The symbolism is similar to that used in Foursquare Figure 9–7, except that a pattern of fine dots has been employed on the vertical surfaces. This was preferable in view of the variable shapes of the French departments. Where the vertical surfaces of adjacent locations meet (and also, on the perimeter of the study space above such meeting places) emphasis has been provided by means of lines within the dot pattern. This contributes to the readability of the map.

As height is increased, those locations with small values become increasingly hidden by those with greater values, as are those departments which comprise Dense France. Map (C), while more dramatic and in some ways the easiest of the three to read, has several locations which are completely hidden.

To simplify matters, classing might be used with this symbolism as shown

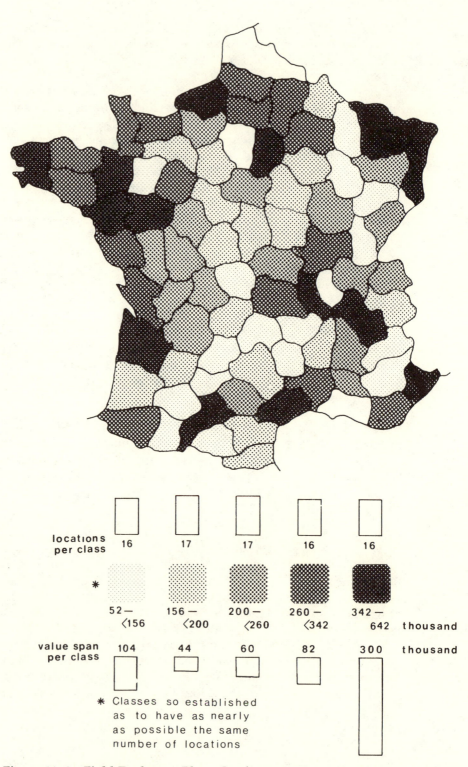

Figure 11–8   Field-Darkness: Plane Conformant, Quantile Classing (8-FD4)

Table 11–2: Quantile Classing for Sparse France

Arrangement	Number of Locations	Unrounded Classes	Rounded Classes
1	16	52,630 – <156,186	52,000 – <156,000
	16	156,186 – <188,633	156,000 – <189,000
	16	188,633 – <254,460	189,000 – <255,000
	17	254,460 – <337,927	255,000 – <340,000
	17	337,927 – 641,917	340,000 – 642,000
2	16	52,630 – <156,186	52,000 – <156,000
	16	156,186 – <188,633	156,000 – <189,000
	17	188,633 – <255,582	189,000 – <255,500
	16	255,581 – <337,927	255,500 – <340,000
	17	337,927 – 641,917	340,000 – 642,000
3	16	52,630 – <156,186	52,000 – <156,000
	17	156,186 – <195,798	156,000 – <200,000
	16	195,798 – <255,581	200,000 – <255,500
	16	255,581 – <337,927	255,500 – <340,000
	17	337,927 – 641,917	340,000 – 642,000
4	17	52,630 – <156,770	52,000 – <156,750
	16	156,770 – <195,798	156,750 – <200,000
	16	195,798 – <255,581	200,000 – <255,500
	16	255,581 – <337,927	255,500 – <340,000
	17	337,927 – 641,917	340,000 – 642,000
5	16	52,630 – <156,186	52,000 – <156,000
	16	156,186 – <188,633	156,000 – <189,000
	17	188,633 – <255,581	189,000 – <255,500
	17	255,581 – <341,485	255,500 – <342,000
	16	341,485 – 641,917	342,000 – 642,000
6	16	52,630 – <156,186	52,000 – <156,000
	17	156,186 – <195,798	156,000 – <200,000
	16	195,798 – <255,581	200,000 – <255,500
	17	255,581 – <341,485	255,500 – <342,000
	16	341,485 – 641,917	342,000 – 642,000
7	17	52,630 – <156,770	52,000 – <156,750
	16	156,770 – <195,798	156,750 – <200,000
	16	195,798 – <255,581	200,000 – <255,500
	17	255,581 – <341,485	255,500 – <342,000
	16	341,485 – 641,917	342,000 – 642,000
8	16	52,630 – <156,186	52,000 – <156,000
	17	156,186 – <195,798	156,000 – <200,000
	17	195,798 – <259,443	200,000 – <260,000
	16	259,443 – <341,485	260,000 – <342,000
	16	341,485 – 641,917	342,000 – 642,000

Table 11–2: Quantile Classing for Sparse France *(continued)*

ARRANGEMENT	NUMBER OF LOCATIONS	UNROUNDED CLASSES	ROUNDED CLASSES
9	17	52,630 – <156,770	52,000 – <156,750
	16	156,770 – <195,798	156,750 – <200,000
	17	195,798 – <259,443	200,000 – <260,000
	16	259,443 – <341,485	260,000 – <342,000
	16	341,485 – 641,917	342,000 – 642,000
10	17	52,630 – <156,770	52,000 – <156,750
	17	156,770 – <203,212	156,750 – <203,500
	16	203,212 – <259,443	203,500 – <260,000
	16	259,443 – <341,485	260,000 – <342,000
	16	341,485 – 641,917	342,000 – 642,000

in Figure 11–11. With only 5 different heights in contrast to 82, the resulting map is far easier to interpret overall. At the same time it is, of course, far less accurate due to the generalizing effect of the classing employed. Since the emphasis here is upon groupings of locations pursuant to the classing used, departmental base outlines have been omitted. In order that the locations may be differentiated, however, since addable values are employed, the assigned base centers of all the departments have been shown by means of dots.

Figure 11–12 is a repetition of the preceding map, but with locations undifferentiated. This requires adjustment in the map's title through the addition of the words "Per Department," indicating the conversion of the values from addable to nonaddable type.

Count   In Figures 11–13A, B and C we see three examples of count symbolism, based on the use of varying element ratings. These correspond to Foursquare Figure 9–8A. There is a different number of classes (8, 16, and 30) in each case. All of the classes, except for the first and last, are of equal span and have well-rounded class limits. With this system, in consequence

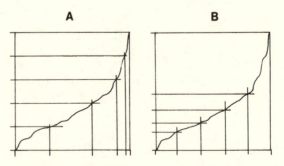

Figure 11–9   Classing Charts for Equal and Quantile Classing

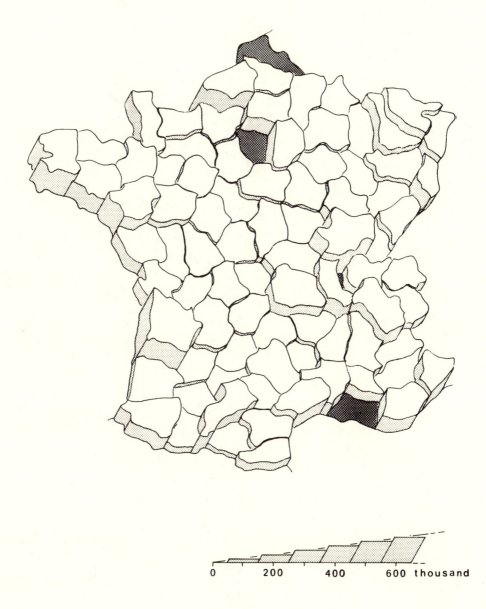

Figure 11–10A   Field-Extent: Raised Conformant (7-FE4), Unclassed, Low Rise

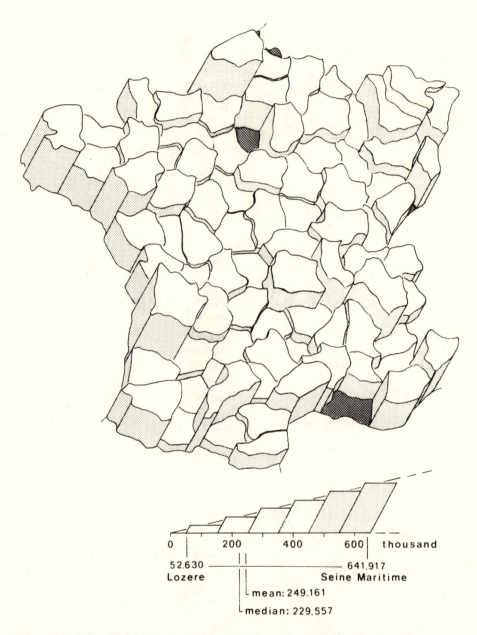

Figure 11–10B   Field-Extent: Raised Conformant (7-FE4), Unclassed, Medium Rise

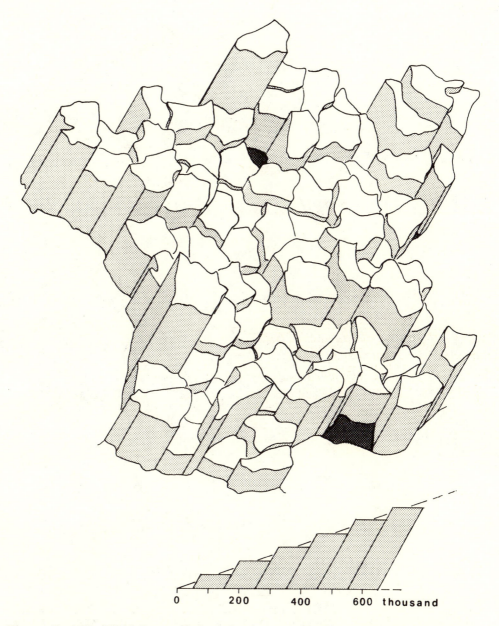

Figure 11–10C   Field-Extent: Raised Conformant (7-FE4), Unclassed, High Rise

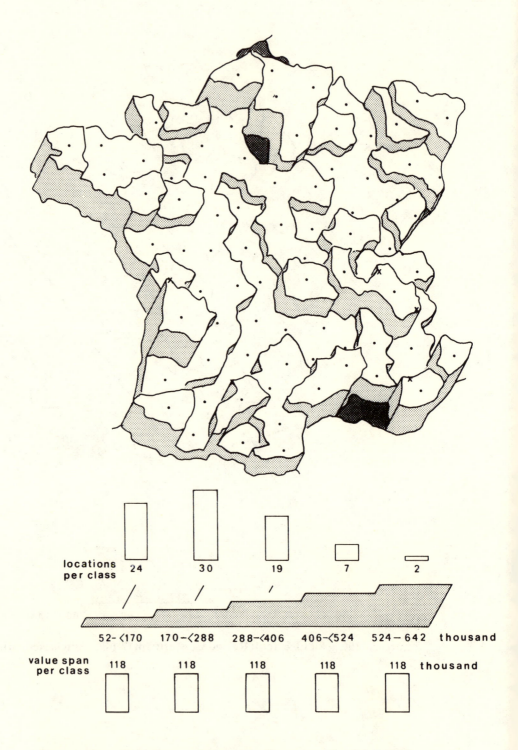

Figure 11–11   Field-Extent: Raised Conformant, Classed (7-FE4)

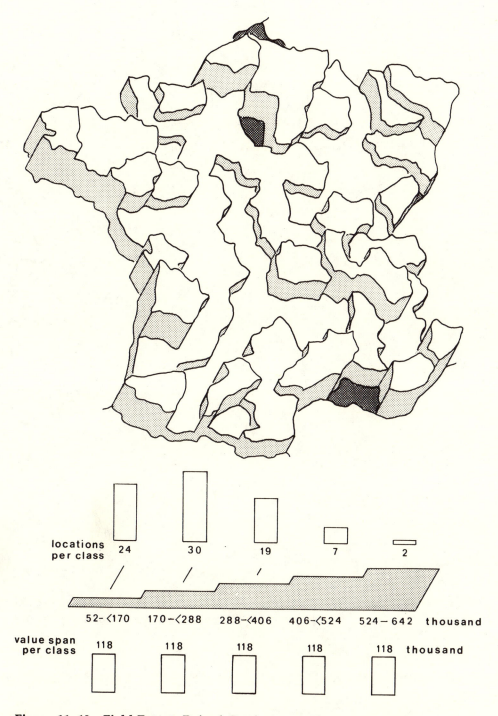

locations
per class 24   30   19   7   2

value span
per class

52-<170   170-<288   288-<406   406-<524   524-642 thousand

118   118   118   118   118 thousand

**Figure 11–12   Field-Extent: Raised Conformant, Classed, Locations
Undifferentiated (7-FE4)**

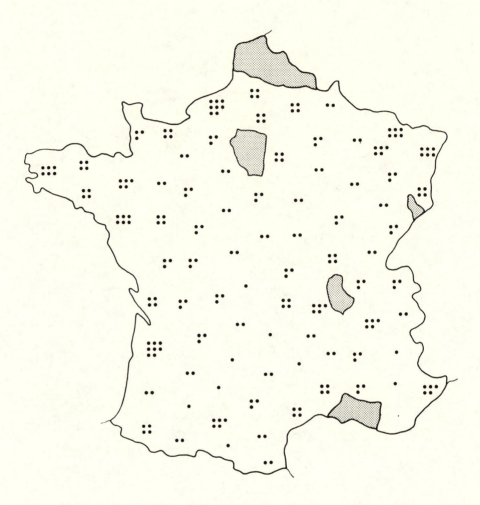

• = 80,000

classed: 1 dot = 52,630 – ⟨120,000
          2 dots = 120,000 – ⟨200,000   etc.

Figure 11–13A   Spot-Count (3-SC3), High Dot Value

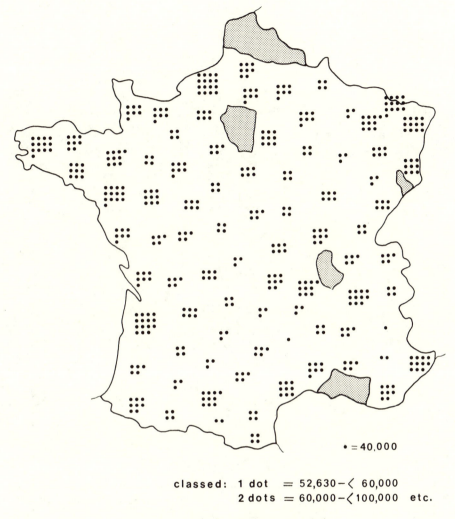

classed: 1 dot = 52,630 – ⟨ 60,000
          2 dots = 60,000 – ⟨100,000 etc.

range: 52,630 – 641,917

Figure 11–13B   Spot-Count (3-SC3), Medium Dot Value

of the large number of classes used and the small class spans involved, the symbols can be read with more than usual accuracy. Only the first two class spans are given in the value key. The others may be easily deduced.

This type of symbolism is especially useful with a rather low value curve. To a considerable degree it combines the advantages of both unclassed and classed symbolism.

It is usually best to have class breaks fall midway between even multiples of the dot rating—as indicated by the illustrative classing examples given in the value key of each map. It is essential that the classing procedure somehow be indicated for those locations where the number of countable elements per symbol is small.

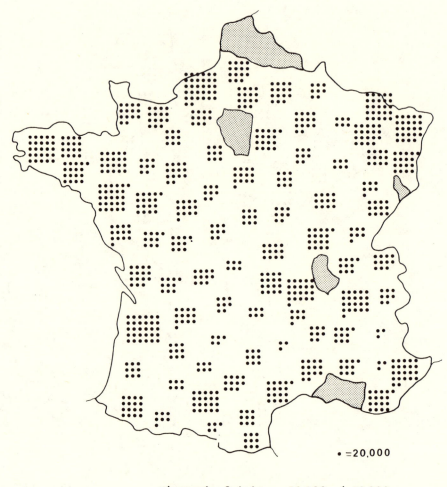

• = 20,000

classed: 3 dots = 52,630 – < 70,000
         4 dots = 70,000 – < 90,000 etc.

Figure 11–13C   Spot-Count (3-SC3), Low Dot Value

With symbolism of this type each symbol must be presented as compactly as possible and the symbols should be visible without overlap. A uniform system for placing the dots within each symbol should be followed. The system here employed is based on a square format insofar as possible, with additional dots being added as required in a single column at the right and then across the bottom until a new larger square is achieved.

Dots   As was discussed earlier, dot mapping is a form of COUNT symbolism in which the dots for each location are spread over the location rather than clustered together at its center in a clearly defined SPOT. Dot maps will usually be read in terms of the relative darknesses, and the dots will seldom actually be counted by the map user. This symbolism is normally used only with addable values—yet it actually represents density, which is nonaddable.

Any of three basic procedures—or a combination of procedures—may be used in establishing the dot positions within each location base outline:

1. The dots may be placed as nearly uniformly as possible.
2. The dots may be placed randomly.
3. The dots may be placed pursuant to special knowledge.

The *first* procedure presents problems in execution. For Sparse France, with locations of dissimilar shape and size, the standard grid employed in Figure 9–9, Foursquare, cannot be used. Reasonably uniform dot spacing may be achieved working by eye (Figure 11–14), but when there are more than a few dots, or the shape of the location is not compact, the task can become quite burdensome.

The *second* procedure—random placement (Figure 11–15)—is now used only with the aid of the computer. Hand production would be so burdensome as to be impracticable. One method employs gridding, usually at a relatively fine scale, within each location—but no attempt is made to achieve uniform spacing. Instead, each dot required within the base outline of a location is positioned in a strictly random manner within one of the location's grid cells, and superimposition of dots is avoided by limiting the closeness of dots within a grid cell. Such a procedure is, of course, *dasymetric* in the extreme— yielding within each location variable dot densities and resulting textures which appear to have meaning, but in fact have none. However, with random placement there is no need to resort to special knowledge, and the difficulties involved in seeking uniform dot spacing are totally avoided.

When this procedure is used, if any portion of a location is incompatible with the subject being represented—and sufficiently large to be significant— it may be left blank. For example, if a portion of a location was swampland and the dots represented people, dots might be entirely omitted there through the use of special information about the location to which the computer program could refer.

A variation of the random placement procedure may be employed, if desired, to yield a more uniform dot distribution and darkness over each base area. This involves gridding at two levels, essentially a stratified random sampling design. Within each location a relatively coarse grid is first employed, after which each grid cell is gridded again at a finer scale. The number of dots required for each location is then divided by the number of coarser grid cells, to establish the number of dots required per cell. This resulting number, with fractions rounded, is then distributed randomly among the finer cells contained within each coarser cell. However, the resulting number of dots per location will not conform precisely with the desired number in consequence of the fractional rounding. This may be an unfortunate consequence in low value areas. In general, unless the map is quite large, this procedure will require the use of very small dots, with the resulting map capable of being read only in terms of unclassed variable darkness.

The *third* procedure, based on the use of special knowledge, is the historic form of dot mapping—to which we will now give our attention (see Figure 11–16). While the placement of the dots is dasymetric, they will normally conform to the true facts far better than when random positioning is used.

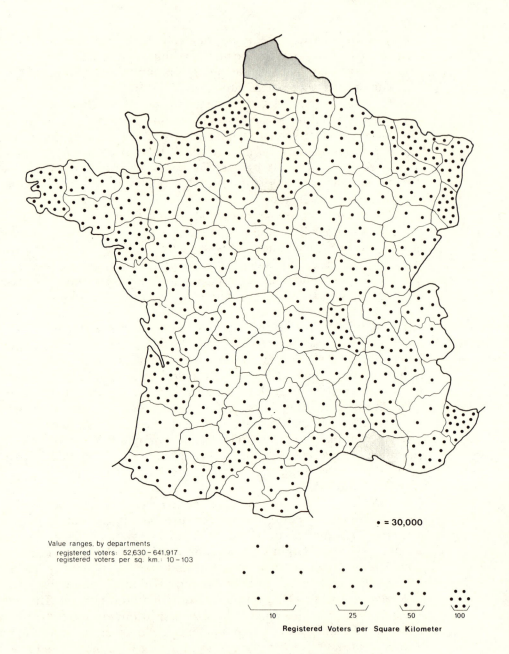

Value ranges, by departments
  registered voters: 52,630 – 641,917
  registered voters per sq. km.: 10 – 103

• = 30,000

10   25   50   100

Registered Voters per Square Kilometer

Figure 11–14   Freehand Uniform Dot Spacing (9-FC5)

TRADITIONAL DOT MAPPING BY HAND   For the production of dasymetric dot
maps of traditional type, it is necessary to employ the somewhat complex
procedure outlined below—though in practice the various steps may in part
at least be subconsciously elided.

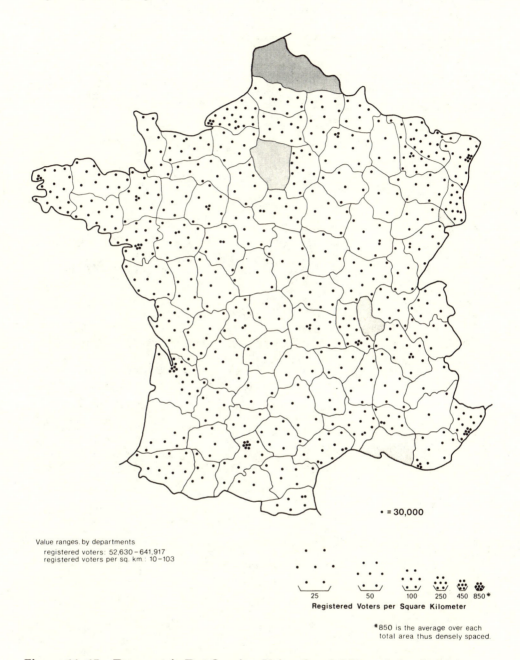

Figure 11–15 **Dasymetric Dot Spacing Using Special Knowledge (9-FC5)**

Step 1. Start with given locations and their relevant values. The values will determine, in *Step 3*, the number of dots to be placed within each location base outline.

Step 2. Any special knowledge regarding the spatial variability of the data within each location, or factors likely to influence such variability, should next be compiled. Library, field, or other re-

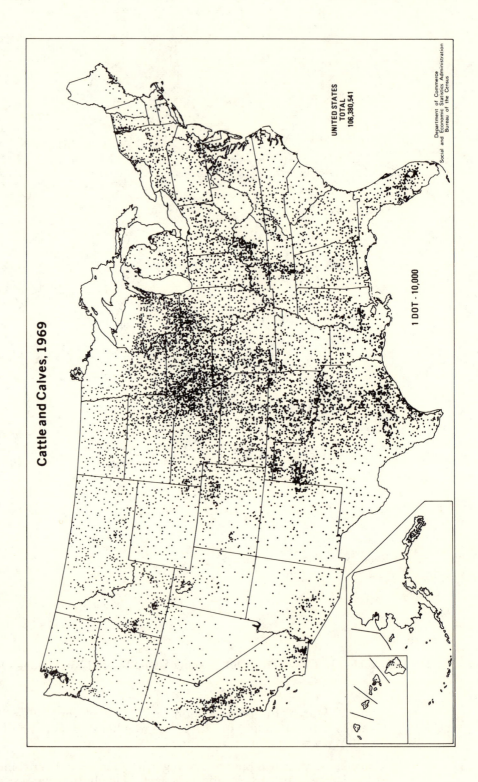

Figure 11–16   Cattle and Calves, 1969

searches should then be undertaken, to the extent warranted and practical. A particular effort should be made to achieve a consistent level of detail over the study space as a whole.

Step 3. Consider suitable dot size and dot "rating." The two goals are densely spaced dots (about to coalesce) in those areas of the map where the highest values exist, and rather sparsely spaced dots in those areas of the map where the lowest values exist. Obviously, in order to determine these, the size of the map must also be established.

Frequently the goals stated above will be impossible to achieve simultaneously. The mapmaker will be faced with having to choose between two undesirable alternatives: having large areas of the map entirely or largely devoid of dots, or using such small dots that they will be visually ineffective and difficult to produce or reproduce. The solution lies in establishing a sufficiently small dot rating to avoid having those areas of the map with low values largely devoid of dots, while using reasonably sized dots and exaggerating as necessary the extent of the high value areas. To avoid exaggerating the high value areas, however, the dot rating should be no smaller than necessary.

Usually at least two or three trials will be made before an optimum approach for the particular circumstances may be achieved. Such trials will involve matters to be considered below.

For any assumed dot rating, the number of dots to be used within a location will be determined by dividing the value for the location by the dot rating. If the result is a fraction, normally the case, the number of dots will depend upon the classing system to be used. As mentioned previously, the most commonly employed procedure and the easiest to comprehend involves the establishment of class breaks at the midpoints between multiples of the dot rating to be used. For example, assuming a dot rating of 1000, three dots would represent the class span of 2500–<3500.

Step 4. In terms of the map size, dot size, and dot rating under *Step 3*, the positioning of the dots applicable to high value areas of the map must next be determined. This will involve the use of special knowledge, and requires judgment. The best procedure will be to place such dots so they almost coalesce. If the map is to be reproduced, consideration must also be given to the likelihood of inked areas enlarging somewhat during the reproduction process. (See Part IV.)

The closely spaced dots in the high value areas of the map will almost always require more space than exists in those areas. Thus, as earlier indicated, the size of the high value areas must be exaggerated. Clusters of densely spaced dots should be centered on and conform in general shape to the high value area represented. However, when a high value area lies wholly within one location, the exaggerated area should be kept wholly within the same location even if its shape must be modified to make that possible. Only thus will the correct number of dots appear within each location.

Step 5. After the positions of all densely spaced dots applicable to high value areas are established, the positions of all remaining dots may be determined. For this purpose, the use of special knowledge and judgment will again be mandatory. The final quality of the map is dependent primarily upon the quality of the special knowledge that is actually employed.

PROBLEMS IN THE USE OF SPECIAL KNOWLEDGE   If special knowledge suggests that the distribution within a given location is approximately uniform, the dots should be placed accordingly. If special knowledge suggests that the distribution is not uniform, the mapmaker must try to place each dot so that it falls in the middle of a compact area which would have a value equal to the dot rating. We will refer to this area as the dot's "tributary area." Each dot might then be thought of as a circle symbol correctly representing the applicable value. If the tributary area must be large in order to have such a value, the resulting dot spacing will be large.

Given a high quality dasymetric dot map of traditional type, it should theoretically be possible to establish a proximal zone centered on each dot which is not densely spaced and have the applicable value conform reasonably closely to the dot rating. To the extent that this is possible, the map will be less than impressionistic. Frequently, however, the maker of the dasymetric dot map will be faced with essentially impossible situations. Imagine, for example, that population is to be mapped and the value for a given location of fair size conforms approximately to the dot rating. If there is a city with about 35 percent or 40 percent of the total population at the west side of the location and a similar city at the east, where should the dot be placed? In traditional dasymetric dot mapping such problems constantly arise. The mapmaker can only attempt what seems most reasonable—or least unreasonable.

For these reasons, cartographers interested in dot mapping and computer use have resorted to the random spacing of dots. A map can be produced that may prove satisfactory when the number of locations is relatively large and hence the area of each location compared to the total study space is relatively small. Thus dot maps of the United States by counties might be employed when such maps by states would be misleading. With counties, the maximum possible error in the placing of any one dot would be limited by the small size of each location in contrast to the larger size of states. While error in relation to counties might be acceptable, error in relation to states would almost certainly not be.

As a practical matter, unless the mapmaker happens to be unusually familiar with both the study space in its entirety and the particular subject to be mapped—which is rarely the case—seriously defective assumptions may be made. In the design of our illustrative map, Figure 11–15, for example, it was first thought likely that the percentages of those of voting age, and hence of those registered, would be greater in more urban areas. In spite of the apparent logic of this analysis, however, the true situation was quite different. The largest percentage of registered voters was in the rural department of Creuse (No. 48), and the smallest in a highly urbanized department adjacent to Paris. Thus even with such a simple subject as registered voters, closely related to overall population (probably the most readily available of all statistical information), it proved necessary to draw upon official census data based upon smaller civil divisions.

When a map deals with a less familiar subject than voters, the problem facing the mapmaker is likely to become much more difficult. Except in rare situations, as when the specialists of the U.S. Department of Agriculture undertake to map agricultural production, the mapmaker cannot possibly become sufficiently expert in each subject to be able to arrive at some judgments as to the intralocational variability present.

The production of the dot map in Figure 11–15 and several others to be presented later was a major effort, often involving several months of work on the part of the author and an experienced cartographer. Yet the resulting maps only roughly suggest the true facts. Although dense clusters of dots are used to represent high value areas, the spaces occupied by these clusters almost always exceed the actual areas of high value. In one example, closely spaced dots represent a density of 850 registered voters per square kilometer (as shown by the value key). However, in the actual high value urban areas, the average density is far greater; in order to accommodate the requisite number of dots at that spacing, the urban areas are necessarily enlarged beyond their actual size.

The basis of our map is registered voters *by departments*. Yet dasymetric dot mapping of historic type shifts the visual emphasis from variability by departments to variability on a much finer scale. For example, in the two easternmost departments of France (No. 8 at the northeast, and No. 63 at the southeast) the intradepartmental variability represented makes it difficult to judge the values applicable to those departments as a whole.

With dasymetric dot mapping of traditional type, adequate citation of sources of special knowledge can be troublesome. For our example the principal sources used were the 82 individual department maps contained in the *Index-Atlas des Departements Français* (Oberthur editeur, Rennes-Paris, 1968) and a map entitled "Parcs Naturels Français" (Institut Géographique National, October 1970, scale 1:1,000,000). With more difficult subjects—such as criminal activity, honey production, or wildlife, for which there may be no clear correlation with easily available statistics—the problem of adequate source citation can frequently be an embarrassment to any conscientious cartographer. Yet a paucity of citation tends to suggest a dearth of valid information.

THE ROLE OF DOT MAPPING   In view of the very large amount of time, thought, and skill required to produce dasymetric dot maps of historic type, and the fact that they are still incapable of withstanding careful scrutiny, the validity and practicality of the system is subject to serious question. As was earlier indicated, since judgment is required for the placing of each separated dot, such maps cannot be made by computer. However, it seems safe to predict that computer-produced dot maps may long continue to prove extremely useful with addable values when the minimum value is a small percentage of the maximum value; when the value curve is low; and when locations are small relative to the total study space. The special value of dot maps under these circumstances results from the fact that these maps will tend to be characterized by a wide range of dot spacing, with resulting darkness tones varying from white to near-black—and with relatively easy differentiation possible among the lighter tones, which will be most prevalent when the value curve is low. The particular combination of circumstances mentioned is frequently encountered and is particularly characteristic of maps of agricultural production. It should be noted that the variable darkness tones are unclassed when locations vary in size or when the dots are placed dasymetrically.

Let us look at an illustration of the usefulness of dot mapping in providing easy differentiability among the tones toward the lighter end of the darkness range. In Sparse France, the department with the least spatial density of registered voters is Lozère (No. 67), with slightly less than ten registered voters per square kilometer. With 52,630 registered voters, Lozère requires only two dots when the rating per dot is 30,000. Yet the difference between the resulting dot spacing (and sense of darkness produced) and that required to represent registered voters per square kilometer, for example, may easily be seen. By reference to the value key swatches of Figure 11–15, a similar comparison can easily be made between the symbolisms for 25, 50, or 100 registered voters per square kilometer. Assuming spatial density values ranging up to 850 registered voters per square kilometer, as represented by the densest swatch of the value key, plane conformant symbolism, in comparison, has very limited usefulness. For example, if we assume the use of five equal classes within the 10–850 value range, the span of the first class would be tremendous, namely 10–<178. Yet all values within that first class would be represented by the same tone of light gray. If we assumed the use of unequal rather than equal classing with conformant symbolism, the lack of differentiation among low values could be somewhat improved but by no means eliminated—and the cost would be a loss of differentiation among high values.

With locations of roughly the same size as those in Sparse France, when the value range for any subject is such that a minimum value is more than a small percentage of the maximum value, dot mapping may not be justified—regardless of the shape of the value curve. Under such circumstances (unless special knowledge is used to produce high value concentrations within locations), all or most of the map would be covered by dots yielding roughly the same average darkness. Instead, plane conformant symbolism could be used with its usual wide darkness range and easy differentiability among the classes represented.

DOT MAPPING WITHOUT LOCATION BASE OUTLINES   With all dot mapping, location base outlines may be omitted as shown in our illustrative dot map of Figure 11–13A. If this map were used, however, the meaning and message of the display would be radically altered, and registered voters by departments would no longer be shown. The title would need to be changed, perhaps to "Sparse France: Registered Voters, Estimated Within Departments—1970," with a brief note of explanation included to make it clear that the raw data used was entirely by departments.

Dot mapping by counties is frequently used when mapping the United States, especially for agricultural data. When so employed, county outlines are usually omitted to avoid overcrowding. State outlines are, however, usually shown (Figure 11–16). There is a good probability, therefore, that the user will make one of two erroneous assumptions: (1) that the basic information was by states, suggesting a far lower level of accuracy than that provided; or (2) that the information was by locations smaller than counties, suggesting a far higher level of accuracy than that provided. One of the disadvantages of dot mapping is that any variability in the spacing of dots tends to suggest corresponding variability in the data. Such problems are not usually encountered with other symbolisms.

When location base outlines are not shown and dot maps are made by computer with random positioning, the display is especially likely to be misleading. Under such circumstances, dots which happen to fall at the edge or corner of one location may combine with similar dots in one or more adjacent locations to create a dark area of concentration. Without the benefit of location base outlines, this is likely to be misinterpreted as representing a high value when in fact it is strictly the result of the random process used. Whether location base areas are shown or not, the user should be informed as to how the map was prepared—otherwise it might erroneously be assumed that particular dot positions have significance.

Under the circumstances mentioned in the preceding paragraph, the wording of the title might perhaps be "Sparse France: Registered Voters—1970," with a note of explanation included stating that the raw data was taken by departments and random dot spacing employed within each department.

Sectors, Fins, and Ribbons   Sector symbolism, as employed in Foursquare Figure 9–10, cannot appropriately be used to represent the registered voters of Sparse France. The use of fin and ribbon symbolisms would be similarly inappropriate, because the departments are irregularly disposed.

Plane Interpolated   The use of areal interpolation, as produced by computer in terms of variable darkness, is shown in Figure 11–17. As in the comparable Foursquare map of Figure 9–13A, each base center is shown here by means of a number corresponding to the number of the class in which the applicable value falls. Class 1 is the class with the lowest value span.

A minor rounding of class spans might well have been used here. It is important to note, however, that whenever contouring or the equivalent is involved, contour positions will normally change when changes are made in the classing breakpoints to which the contour line ratings conform.

Since the values used are addable, they only have meaning in terms of the locations involved, here represented by numerals placed at base centers. This type of symbolism, however, can be of value as a guide to the general trends of the values.

The validity of employing interpolation with addable values when locations are differentiated is further demonstrated by Figures 11–18A, B and C. The first map, Map A, is meaningful but difficult to comprehend. Map

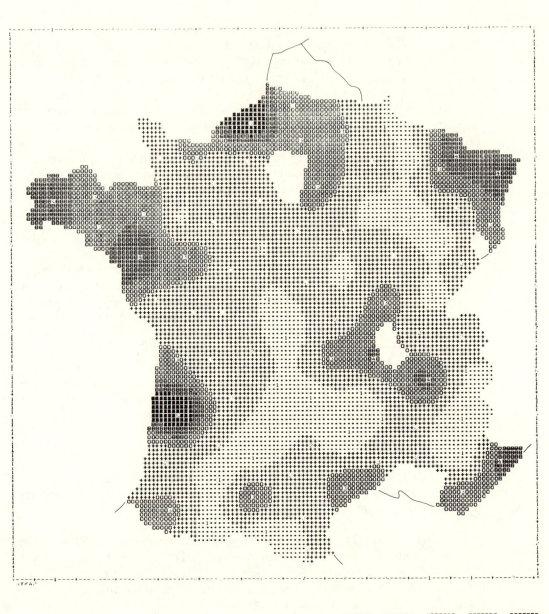

| 52.630–
<170.487 | 170.487–
<288.345 | 288.345–
<406.202 | 406.202–
<524.059 | 524.059–
641.917 |

Figure 11–17   Field-Darkness: Plane Interpolated, with Variable Darkness (8-FD7)

B adds contour lines of graphically readable type, and the map is far more easily understood. Map C uses dots rather than numerals at the location base centers. While the message here is perhaps less obvious, it is still far

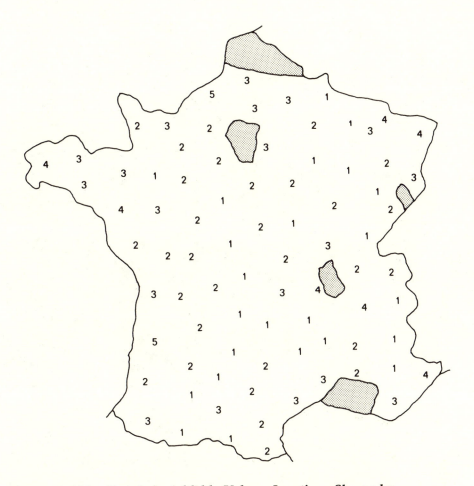

Figure 11–18A Trends in Addable Values, Locations Shown by Class Number

more easily read than Map A. The precise positioning of the contour lines serves to give some idea of the actual unrounded values involved. In contrast, the numerals alone give no clue as to the actual values represented by the class spans.

Figure 11–19 shows a hand-drawn map comparable to the above computer map, except that base centers have been omitted and contour lines have been drawn between contiguous tones—as in Foursquare Figure 9–15F. Since it has been traced over a reduction of the preceding computer map, the study space perimeter has been substantially simplified compared with other hand-drawn maps of Sparse France. This minimizes certain complexities which might be confusing in relation to the contour lines—most notably along the west-central coastline, facing the Bay of Biscay. In other interpolated maps of France to follow, this same generalization will be used.

Compare this map with the conformant map of Figure 11–6. The values, the classing procedure, and the darkness tones are identical in both. As-

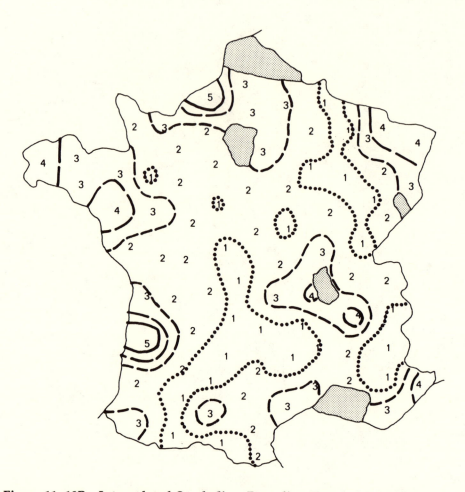

Figure 11–18B   Interpolated Symbolism Revealing Trends in Addable Values, Graphic Contour Lines and Class Numbers Shown

suming that primary concern is with the values of the individual locations—their shapes and base outline relationships—it is obvious that the interpolated version of Figure 11–19 will not serve. If, however, primary concern is with the trend of the values over the study space, the interpolated version will undoubtedly prove far more revealing. If the locations in the interpolated version had been differentiated, the two maps would be equally effective in revealing the class in which each value falls. The choice between the conformant and interpolated maps would then be between a display hard to interpret, but showing each departmental base outline, and one relatively easy to interpret, but not showing each departmental base outline.

When interpolation is used, any areas internal to but not a part of the study space, tend to interrupt the continuity of the darkness tones as well as any contour lines separating them. In the present instance, the areas of Dense France create no significant difficulty. In mapping urban housing conditions by blocks, however, it would be far better to carry the contour lines across all intervening streets than to interrupt them.

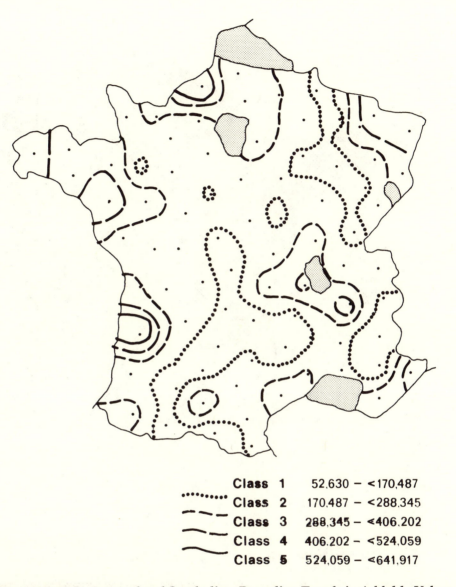

	Class	1	52.630 – <170.487
••••••••••	Class	2	170.487 – <288.345
– – –	Class	3	288.345 – <406.202
——	Class	4	406.202 – <524.059
——	Class	5	524.059 – <641.917

Figure 11–18C   Interpolated Symbolism Revealing Trends in Addable Values, with Graphic Contour Lines and Legend

An interpolated map similar to the last several presented but with classing according to the quantile method might be employed. Its classing would correspond to the conformant map of Figure 11–8, assuming the moderate rounding earlier suggested. With such a map, the contour ratings, as compared to those for five equal classes, would be as follows: 156 (vs. 170), 200 (vs. 288), 260 (vs. 406), and 342 (vs. 524). As with the conformant map of Figure 11–8, the adjustments shown would serve to place (as nearly as possible) the same number of locations in each class.

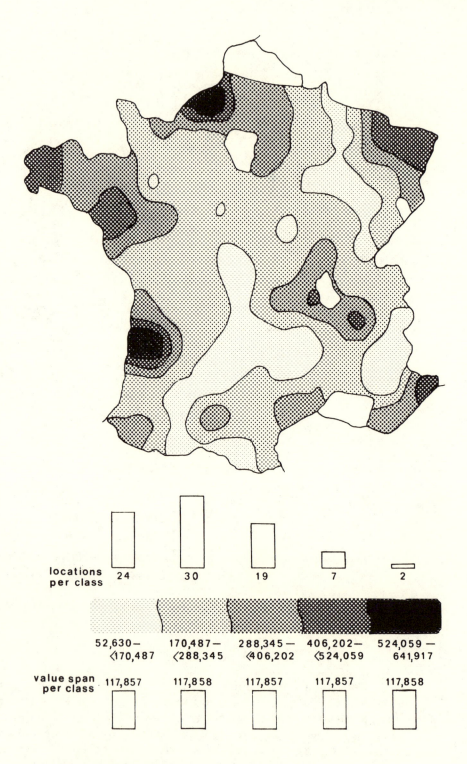

Figure 11–19   Hand-Drawn Version of Figure 11–17, with Variable
Darkness Added and Base Centers Omitted

Figure 11–20 shows an interpolated map with the classes differentiated by shadow contours.

Raised Interpolated Figures 11–21A, B and C show raised counterparts of the several preceding interpolated maps, completely unclassed. These were computer produced (here with a small amount of handwork added). At each location base center, here undifferentiated in view of the overall size of the map, the height of the conceptual surface is proportional to the applicable value. Everywhere else over the entire study space, height is proportional to values established by interpolation.

In Map A the entire conceptual surface is visible. In the other two maps

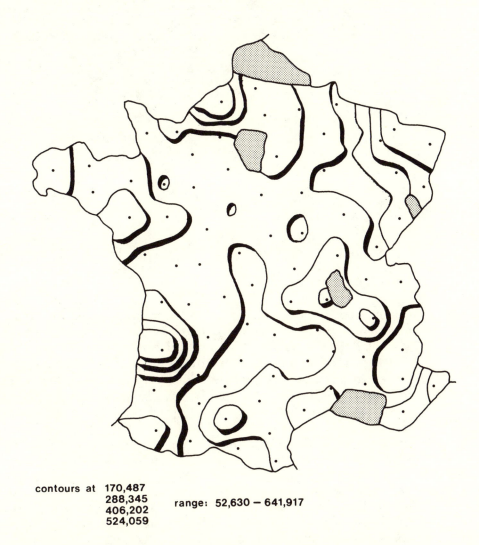

contours at 170,487
288,345   range: 52,630 – 641,917
406,202
524,059

Figure 11–20   Interpolation with Shaded Contours, Base Centers Differentiated

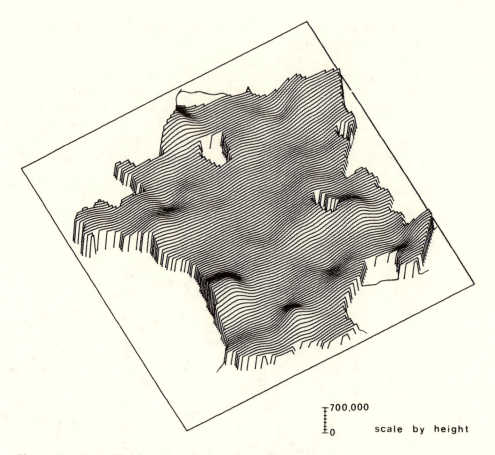

Figure 11–21A   Field-Extent: Raised Interpolated (7-FE6), Unclassed, Low Rise

some portions of the surface are hidden from view, but the greater heights used tend to provide easier readability. When hidden portions of a surface are not extensive, as here, the adjacent curvatures of the surface-defining lines will usually provide a good clue to what is hidden. (Location base centers may be shown on maps of this kind by asterisks, but would tend to be confusing with maps as small as those here presented.)

In view of the oblique disposition of the study space in these maps, and the consequent foreshortening, it is desirable to have a map border to help the viewer. Maps directly corresponding to these can, of course, be produced without foreshortening.

RAISED INTERPOLATED, CONTOURED   In Figures 11–22A, B and C, corresponding to Foursquare Figure 9–16A, the virtues of raised, totally unclassed interpolation and classing by contouring are combined. The contour ratings used here are based on five equal classes between the extremes, rounded only enough to avoid fractions. Showing location base centers here would have complicated the picture even more than with the maps of the preceding

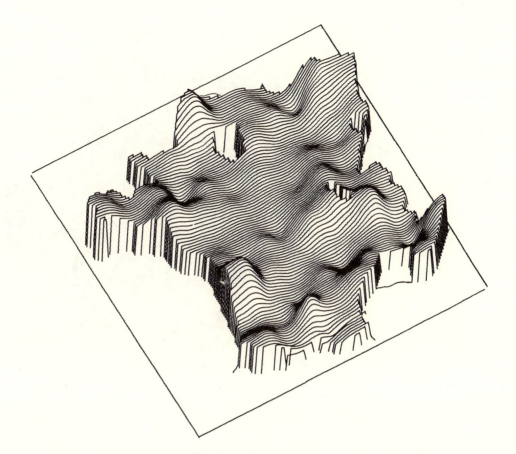

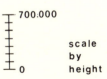

scale
by
height

range: 52,630 — 641,917

Figure 11–21B   Field-Extent: Raised Interpolated (7-FE6), Unclassed, Medium Rise

700.000

scale
by
height

0

Figure 11–21C   Field-Extent: Raised Interpolated (7-FE6), Unclassed, High Rise

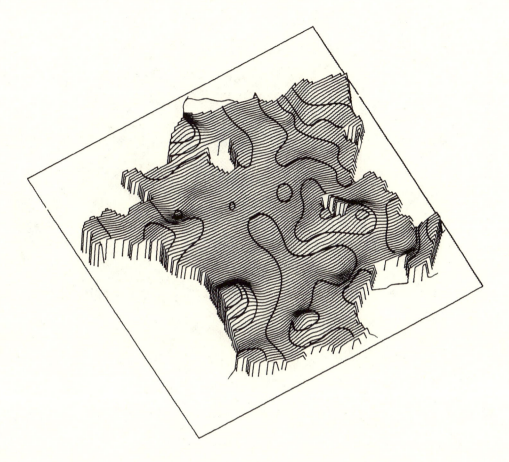

700,000

0

scale by height

Figure 11–22A   Field-Extent: Raised Interpolated, Unclassed in Height, Classed by Contours, Low Rise

contours at 170,487
                    288,345
                    406,202
                    524,059

range: 52,630 — 641,917

700.000

scale
by
height

0

**Figure 11–22B   Field-Extent: Raised Interpolated, Unclassed in Height,
Classed by Contours, Medium Rise**

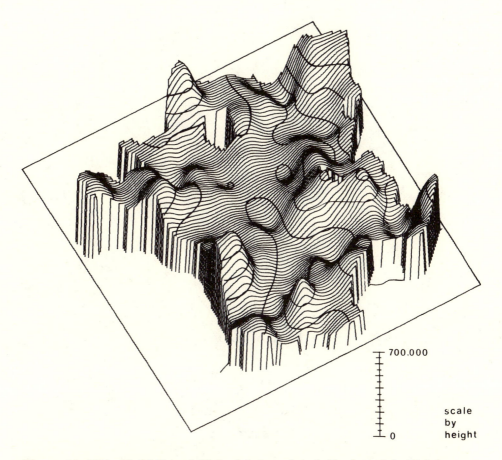

700.000

scale
by
height

0

Figure 11–22C   Field-Extent: Raised Interpolated, Unclassed in Height, Classed by Contours, High Rise

figure. At substantially larger scale, however, location base centers could have been added with no difficulty.

RAISED INTERPOLATED, STEPPED   Figures 11–23A, B and C correspond to Foursquare Figure 9–17A, except that locations have been differentiated by dots placed at their base centers. Again, moderately rounded class span limits might well have been desirable. However, in spite of the extremely ragged values without rounding, there is great merit in the concept of equal classing within the *actual* value range involved.

By comparing this symbolism with that of Figure 11–10, we can see clearly the advantages of interpolation. Not only are the more smoothly flowing contour lines far easier to grasp, but the various levels are always sequential. In contrast, with the conformant symbolism of Figure 11–10, adjoining levels are in some cases as much as three classes apart.

The assumed viewing position has been offset slightly toward the lower left, rather than being directly overhead, making vertical surfaces toward the viewer visible. These maps might have been viewed from an angle and

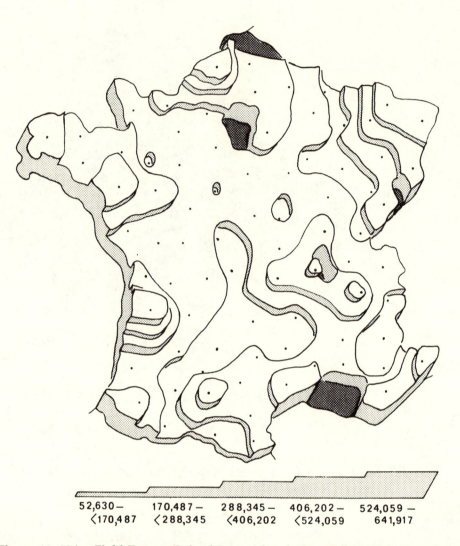

52,630 —	170,487 —	288,345 —	406,202 —	524,059 —
⟨170,487	⟨288,345	⟨406,202	⟨524,059	641,917

Figure 11–23A   Field-Extent: Raised Interpolated, Stepped (7-FE7), Low Step

foreshortened like Figures 11–21 and 11–22. (Or the maps of Figures 11–21 and 11–22 might have been presented unforeshortened as here.)

We have now illustrated for Sparse France most of the graphic symbolisms originally presented for Foursquare, though by no means all possible variations and combinations.

Collaborative Symbolisms   Before leaving Sparse France, let us look at two examples of collaborative symbolisms. Figure 11–24 is similar to Figure 11–23B, except that location base centers have been omitted and variable tones added—as suggested by Foursquare Figure 9–22. With this combination of symbolisms it would be difficult to show the base centers without substantially enlarging the map. General readability has certainly been improved. Figure 11–25 presents another example of the collaborative approach. Un-

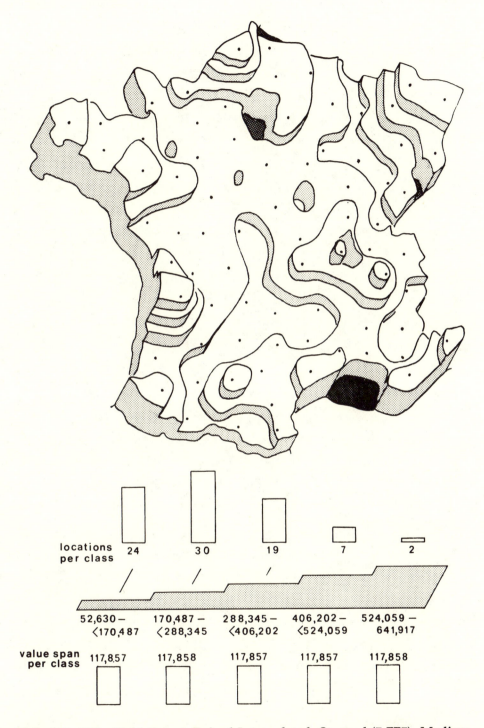

Figure 11–23B   Field-Extent: Raised Interpolated, Stepped (7-FE7), Medium Step

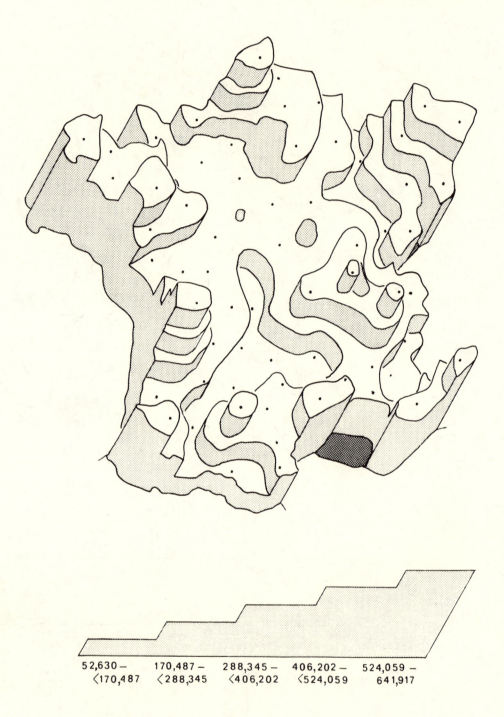

| 52,630 — | 170,487 — | 288,345 — | 406,202 — | 524,059 — |
| ⟨170,487 | ⟨288,345 | ⟨406,202 | ⟨524,059 | 641,917 |

Figure 11–23C   Field-Extent: Raised Interpolated, Stepped (7-FE7), High Step

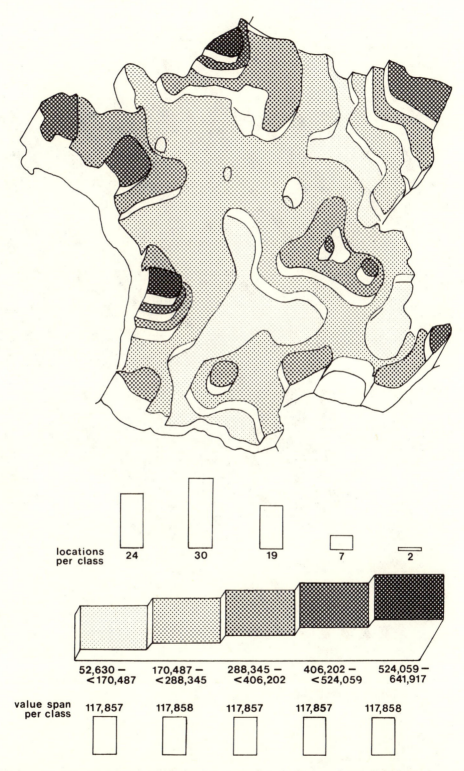

Figure 11–24   Field-Extent: Raised Interpolated, Stepped, Variable Darkness Added and Base Centers Omitted

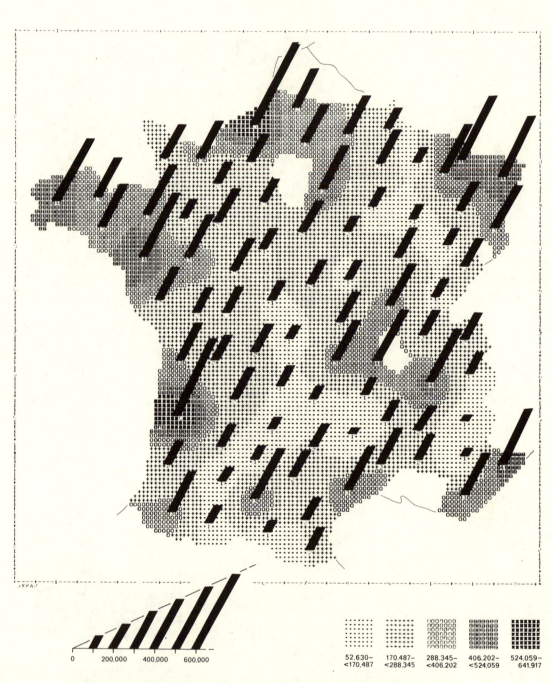

**Figure 11–25   Collaborative Symbolism with Interpolation and
Superimposed Bars**

classed bars convey the number of registered voters in each department, and plane interpolated symbolism has been added. The effect of grouping the bars into five equal classes (represented by five different levels of darkness) is that overall comprehension is improved as compared to Figure 11–5.

DENSE FRANCE

Editor's note: Although extant text for the following section is very limited, the illustrations further the inquiry begun with Sparse France: the investigation of symbolism in a real-world situation. The reader is invited to make comparisons among the figures, referring back to Foursquare and Sparse France for detailed discussions of symbolisms.

The Study Space

Our concern in this discussion is with that portion of coterminous France which has a population density of at least 500 persons per square mile. For convenience, we will refer to this as Dense France.

The study space for Dense France consists of five separate regions. The configuration of these regions will present problems in terms of the symbolization process, since the predominant high values are clustered in some areas and thinly spread elsewhere.

The Locations and Values

Again we are dealing with the number of voters that were registered in each department on February 28, 1970.

Figure 11–26 shows the procedure used in establishing the reference number employed to identify the various locations, with the reference sequence confirmed to avoid any possible confusion between the locations of Sparse France and Dense France.

In contrast to the 82 locations of Sparse France, each with less than 500 persons per square mile, we now have only 12 locations, each with 500 or more persons per square mile, as previously shown within the gray area of Figure 11–1. The location base areas of Dense France vary greatly in size, as can be seen in Figure 11–27 comparable to Sparse France Figure 11–2. Note that the smallest department (Ville de Paris) is 95 percent smaller than the mean of all the departments, and that the largest department is 191 percent larger than the mean. Thus the smallest department is about one-twentieth of the mean size, while the largest has a little more than 38 times the area of the smallest. Compare this with Sparse France, where the largest department was only 3.7 times greater than the smallest. These differences are vast—a fact which will present problems in terms of the symbolization process.

As with Sparse France, the value of each location could be assigned to its base center.

The Value Set Table 11–3 provides the information which is to be portrayed for each location within Dense France.

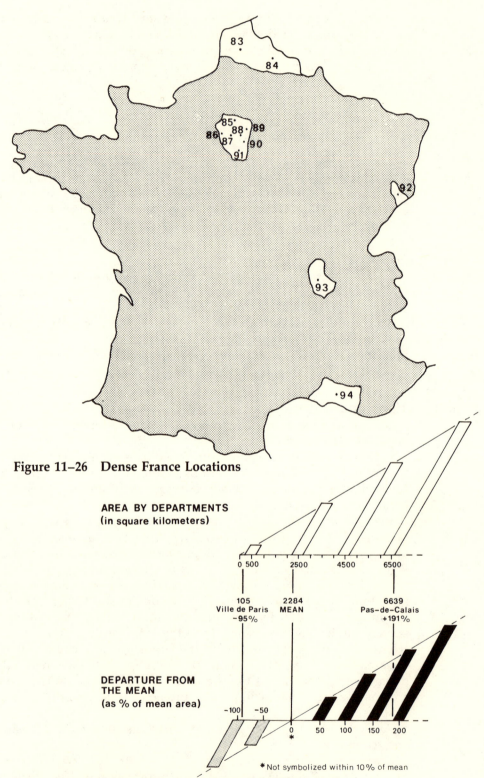

Figure 11–26   Dense France Locations

Figure 11–27   Value Key: Symbolic Representation of Department Areas in Dense France (map on facing page)

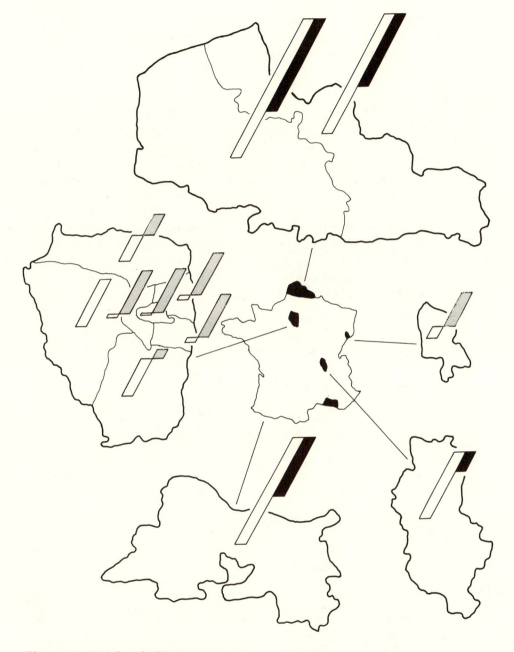

Figure 11–27   Symbolic Representation of Department Areas in Dense France (value key on facing page)

The Illustrations   Figures 11–28 through 11–34 include the value curve and representative samples of how Dense France can be mapped with various symbolisms to show registered voters.

ALL FRANCE

Table 11–4 and Figures 11–35 through 11–45 simply add the Dense France values to those of Sparse France, to map the country as All France I. Some special classing procedures are used. Figures 11–46 through 11–51B combine the values from Sparse France and Dense France in a per square kilometer form to show the density of registered voters as All France II. The conversion to density values overcomes the extreme variation in the actual size of the departments by reference to a common unit of measurement. The problem of symbolism is the representation of a wide range of values distributed in a very low value curve. While the various classing methods are serviceable, equal classing should be avoided for the reasons outlined in Chapter 10, the section on value positions and value curves.

Table 11–3: Dense France: Registered Voters, by Departments—1970

	Increasing Value Sequence	
MAP REFERENCE NOS.	LOCATIONS	VALUES
92	Territoire de Belfort*	64,901
85	Val d'Oise	371,178
91	Essonne	371,415
86	Yvelines	470,965
90	Val-de-Marne	600,808
89	Seine-Saint-Denis	633,599
93	Rhône	693,934
94	Bouches-du-Rhône	772,984
83	Pas-de-Calais	775,942
87	Hauts-de-Seine	796,461
84	Nord	1,343,515
88	Ville de Paris	1,423,528
	TOTAL	8,319,230

*Treated as a department.

Source: Institut National de la Statistique et des Etudes Economiques, Annuaire Statistique de la France 1972, 77ᵉ Volume, Resultats de 1970, Nouvelle Serie No. 19, Ministère de l'Economie et des Finances, Republique Française, p. 125.

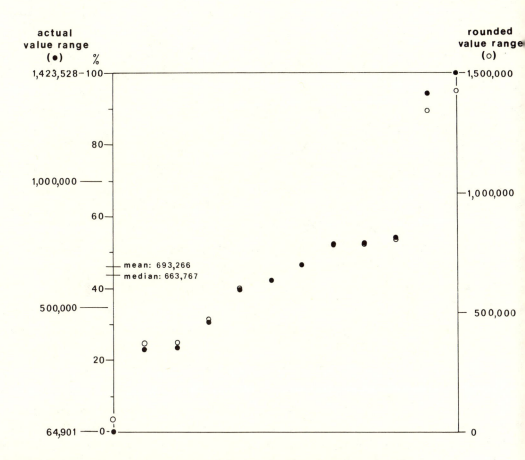

Figure 11–28 The Dense France Value Curve, Showing Actual and Rounded Ranges

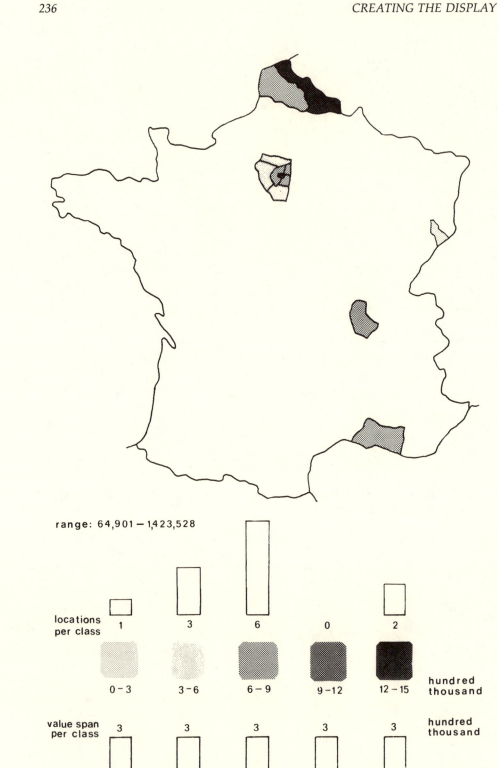

Figure 11–29   Field-Darkness: Plane Conformant (8-FD4), Equal Classing

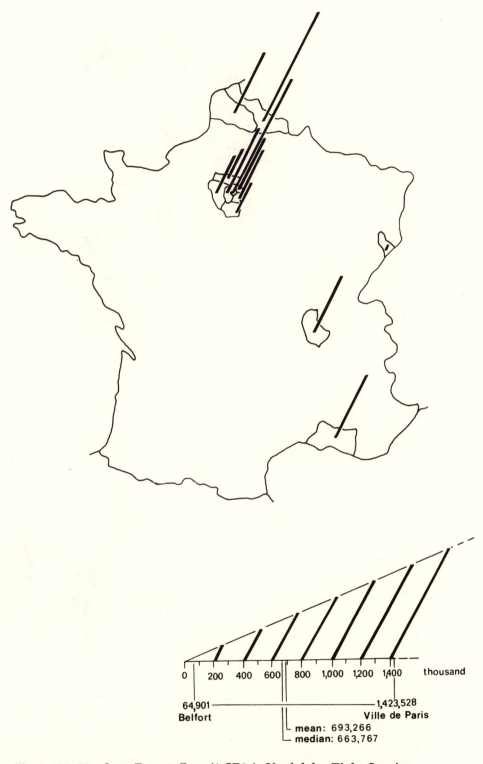

Figure 11–30   Spot-Extent: Bars (1-SE1a), Useful for Tight Spacing

Figure 11–31   Spot-Extent: Sectors (1-SE4a), Judged by Area

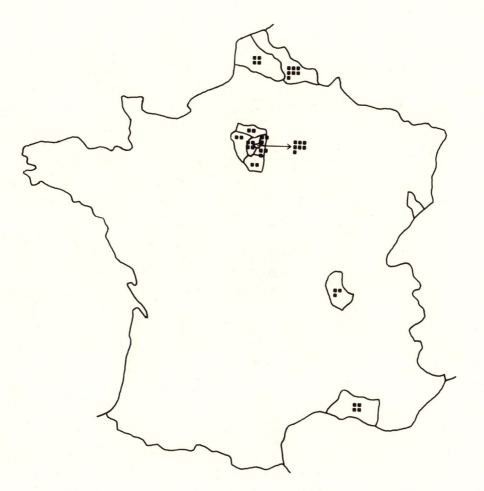

range: 64,901 — 1,423,528

• = 200,000
<100,000 not shown

Figure 11–32   Spot-Count (3-SC2) (Note Forced Displacement of Symbol for Ville de Paris)

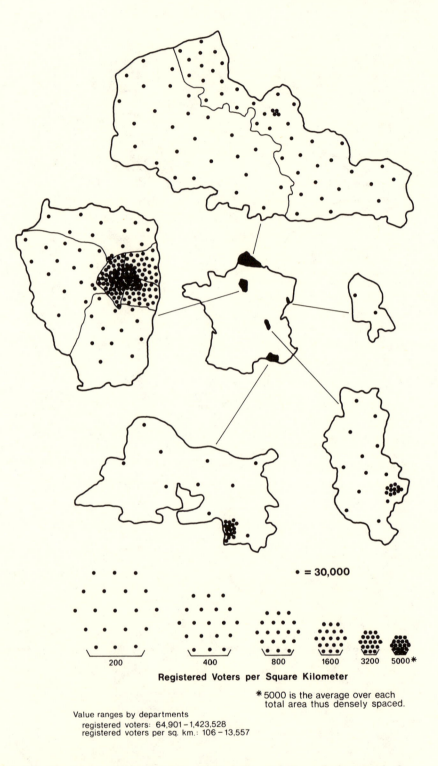

• = 30,000

Registered Voters per Square Kilometer

200   400   800   1600   3200   5000 *

*5000 is the average over each
 total area thus densely spaced.

Value ranges by departments
  registered voters: 64,901 – 1,423,528
  registered voters per sq. km.: 106 – 13,557

**Figure 11–33   Dasymetric Dot Map Using Special Knowledge; Includes
Density Value Key**

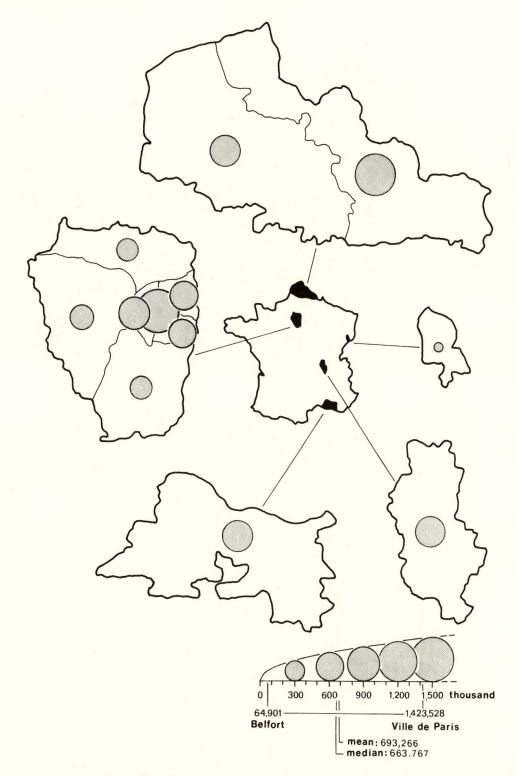

Figure 11–34 Spot-Extent: Circles (1-SE3)

Table 11–4: All France I Value Set
With Break Points for Quantile Classing

Map Ref. Nos.[1]	By Departments	Accum.	By Area (sq. km.)	Accum.	By Registered Voters	Accum.
67	1	1	5,168	5,168	52,630	52,630
61	1	2	5,520	10,688	59,252	111,882
92	1	3	610	11,298	64,901	176,783
62	1	4	6,944	18,242	66,102	242,885
79	1	5	4,890	23,132	94,813	337,698
69	1	6	5,228	28,360	101,956	439,654
57	1	7	5,741	34,101	110,484	550,138
48	1	8	5,559	39,660	110,546	660,684
74	1	9	3,716	43,376	113,700	774,384
73	1	10	6,254	49,630	115,314	889,698
5	1	11	6,220	55,850	121,693	1,011,391
10	1	12	6,216	62,066	122,751	1,134,142
23	1	13	5,343	67,409	132,997	1,267,139
58	1	14	4,965	72,374	137,406	1,404,545
81	1	15	4,507	76,881	142,043	1,546,588
26	1	16	5,008	81,889	145,036	1,691,624
32	1	17	5,171	87,060	155,626	1,847,250
28	1	18	6,837	93,897	156,746	2,003,996
11	1	19	6,002	99,899	156,794	2,160,790
40	a 1	20	a 6,778	106,677	162,220	2,323,010
56	1	21	5,860	112,537	162,400	2,485,410
4	1	22	5,219	117,756	166,277	2,651,687
30	1	23	6,314	124,070	166,810	2,818,497
59	1	24	5,523	129,593	167,377	2,985,874
44	1	25	6,036	135,629	167,400	3,153,274
18	1	26	6,100	141,729	171,872	3,325,146
78	1	27	4,086	145,815	172,014	3,497,160

Rank		Index				
21	1	28	7,425	153,240	176,660	3,673,820
77	1	29	6,232	159,472	177,165	3,850,985
70	1	30	5,358	164,830	179,068	4,030,053
19	1	31	5,876	170,706	181,595	4,211,648
72	1	32	9,237	179,943	181,949	4,393,597
68	1	33	8,735	188,678	188,174	4,581,771
29	1	34	7,228	195,906	189,092	4,770,863
51	1	35	6,004	201,910	202,503	4,973,366
42	1	36	5,756	207,666	203,921	5,177,287
60	1	37^2	6,525	214,191	205,134	5,382,421
			b	b		
65	1	38	3,566	217,757	206,440	5,588,861
54	1	39	5,953	223,710	206,730	5,795,591
					a	
50	1	40	6,985	230,695	209,192	6,004,783
75	1	41	5,751	236,446	211,813	6,216,596
43	1	42	4,391	240,837	216,983	6,433,579
15	1	43	6,004	246,841	224,130	6,657,709
49	1	44	5,512	252,353	229,605	6,887,314
25	1	45	5,228	257,581	230,133	7,117,447
9	1	46	5,871	263,452	233,955	7,351,402
22	1	47	8,765	272,217	239,674	7,591,076
41	1	48	7,327	279,544	243,702	7,834,778
55	1	49	9,184	288,728	253,424	8,088,202
20	1	50	6,742	295,470	255,496	8,343,698
39	1	51	6,124	301,594	255,665	8,599,363
52	1	52	6,721	308,315	263,220	8,862,583
12	1	53	8,163	316,478	270,150	9,132,733
17	1	54	5,947	322,425	271,360	9,404,093
			c			
31	1	55	6,210	328,635	275,228	9,679,321
66	1	56	5,848	334,483	288,510	9,967,831
	c	c				
16	1	57	5,536	340,019	296,508	10,264,339
3	1	58	7,378	347,397	296,978	10,561,317
53	1	59	6,848	354,245	298,138	10,859,455
14	1	60	5,857	360,102	298,322	11,157,777

Table 11–4: All France I Value Set
With Break Points for Quantile Classing (*continued*)

Map Ref. Nos.[1]	By Departments Accum.	By Area (sq. km.)	Accum.	By Registered Voters	Accum.
2	61	6,175	366,277	b 304,455	11,462,232
34	62	6,878	373,155	314,440	11,776,672
82	63	7,629	380,784	318,115	12,094,787
47	64	7,955	388,739	324,639	12,419,426
27	65	8,565	397,304	334,994	12,754,420
64	66	5,999	403,303	335,288	13,089,708
36	67	6,763	410,066	340,566	13,430,274
38	68	7,131	417,197	342,404	13,772,678
76	69	6,113	423,310	342,615	14,115,293
24	70	d 3,523	426,833	345,170	14,460,463
13	71	5,917	432,750	348,184	14,808,647
85	72	1,249	433,999	371,178	15,179,825
91	73	1,811	435,810	371,415	15,551,240
6	74	5,235	441,045	380,962	15,932,202
33	d 75	6,758	447,803	396,852	16,329,054
80	76	6,301	454,104	402,225	16,731,279
46	77	4,774	458,878	c 417,998	17,149,277
45	78	7,474	466,352	424,589	17,573,866
63	79	4,294	470,646	433,734	18,007,600
86	80	2,271	472,917	470,965	18,478,565
8	81	4,787	477,704	472,740	18,951,305
7	82	6,214	483,918	500,678	19,451,983
35	83	6,785	490,703	502,828	19,954,811
37	84	6,893	497,596	510,750	20,465,561
90	85	244	497,840	600,808	21,066,369
71	86	10,000	507,840	602,173	21,668,542
89	87	236	508,076	633,599	22,302,141
1	88	6,254	514,330	d 641,917	22,944,058

By Departments			By Area (sq. km.)		By Registered Voters	
93	1	89	3,215	517,545	693,934	23,637,992
94	1	90	5,112	522,657	772,984	24,410,976
83	1	91	6,639	529,296	775,942	25,186,918
87	1	92	175	529,471	796,461	25,983,379
84	1	93	5,738	535,209	1,343,515	27,326,894
88	1	94	105	535,314	1,423,528	28,750,422

	By Departments	By Area (sq. km.)	By Registered Voters
Desired Quintile Breakpoints	a 18.8	a 107,062.8	a 5,750,084.4
	b 37.6	b 214,125.6	b 11,500,168.8
	c 56.4	c 321,188.4	c 17,250,253.2
	d 75.2	d 428,251.2	d 23,000,337.6
	(94)	(535,314)	(28,750,422)
Closest Possible Quintile Breakpoints[3]	a 19	a 106,677	a 5,795,591
	b 37	b 214,191	b 11,462,232
	c 56	c 322,425	c 17,149,277
	d 75	d 426,833	d 22,944,058
	(94)	(535,314)	(28,750,422)
Desired Quintile Span Per Class	18.8 (20%)	107,062.8 (20%)	5,750,084.4 (20%)
Closest Possible Quintile Span Per Class	19 (20.21%)	106,677 (19.93%)	5,795,591 (20.16%)
	18 (19.15%)	107,514 (20.08%)	5,666,641 (19.71%) min.
	19 (20.21%)	108,234 (20.22%)	5,687,045 (19.78%)
	19 (20.21%)	104,408 (19.50%) min.	5,794,781 (20.16%)
	19 (20.21%)	108,481 (20.26%) max.	5,806,364 (20.20%) max.

Table 11–4: All France I Value Set
With Break Points for Quantile Classing (continued)

	By Departments	By Area (Sq. Km.)	By Registered Voters
Value Span Per Class[4]			
	52,630 – <159,507	52,630 – <162,310	52,630 – <207,961
	159,507 – <205,787	162,310 – <205,787	207,961 – <309,447.5
	205,787 – <292,509	205,787 – <273,294	309,447.5 – <421,293.5
	292,509 – <399,538.5	273,294 – <346,677	421,293.5 – <667,925.5
	399,538.5 – 1,423,528	346,677 – 1,423,528	667,925.5 – 1,423,528
Rounded Value Span Per Class[5]			
	52,000 – <160,000	52,000 – <162,300	52,000 – <208,000
	160,000 – <206,000	162,300 – <206,000	208,000 – <309,000
	206,000 – <293,000	206,000 – <273,000	309,000 – <421,000
	293,000 – <400,000	273,000 – <347,000	421,000 – <668,000
	400,000 – 1,424,000	347,000 – 1,424,000	668,000 – 1,424,000

[1] Ranked from small to large in terms of registered voters.
[2] See Note 5.
[3] Necessary to avoid breakpoints which would fragment locations.
[4] For "By Departments", the value of each breakpoint was obtained by taking the midpoint between the two adjacent given values.
[5] Value spans rounded to the nearest 1000—or, when impossible, to the nearest 100. (For "By Departments", placing the smaller number of locations, 18, in the second class permitted rounding to the nearest 1000.)

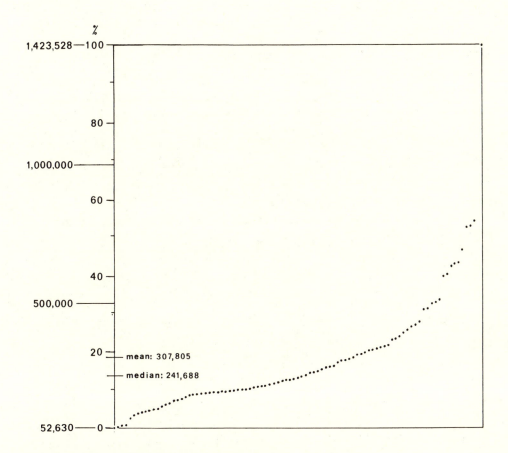

Figure 11–35   The All France I Value Curve

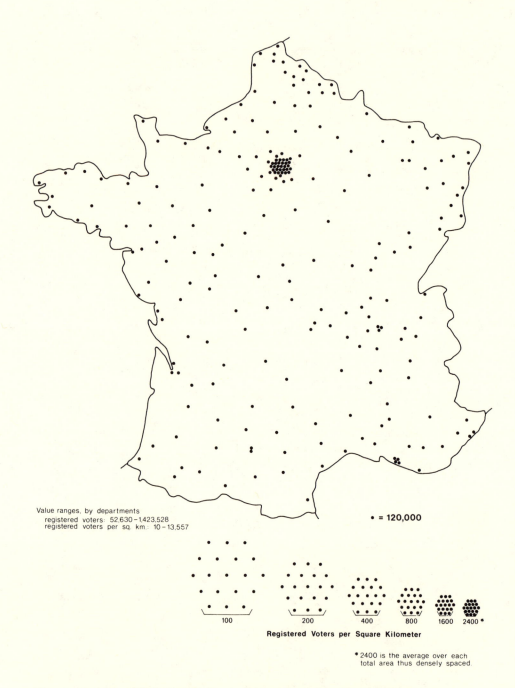

Value ranges, by departments
  registered voters: 52,630 – 1,423,528
  registered voters per sq. km.: 10 – 13,557

• = 120,000

Registered Voters per Square Kilometer

100   200   400   800   1600   2400 *

* 2400 is the average over each
  total area thus densely spaced.

Figure 11–36A   High Dot Value, without Location Boundaries

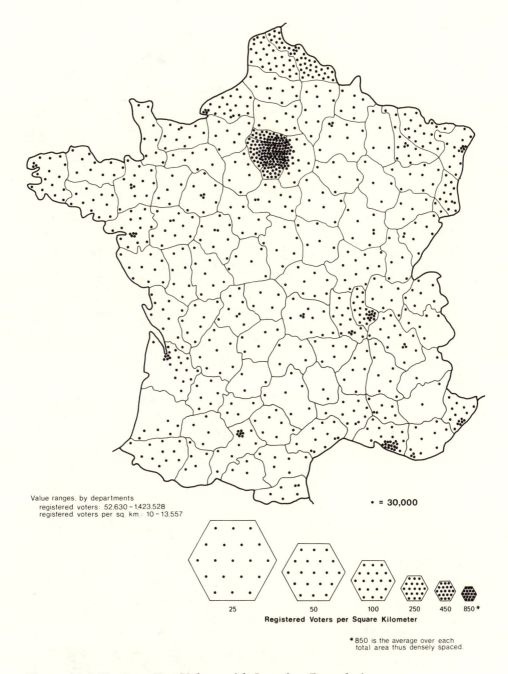

Value ranges, by departments
  registered voters: 52,630 – 1,423,528
  registered voters per sq. km.: 10 – 13,557

• = 30,000

25   50   100   250   450   850*

Registered Voters per Square Kilometer

*850 is the average over each
 total area thus densely spaced.

Figure 11–36B   Low Dot Value, with Location Boundaries

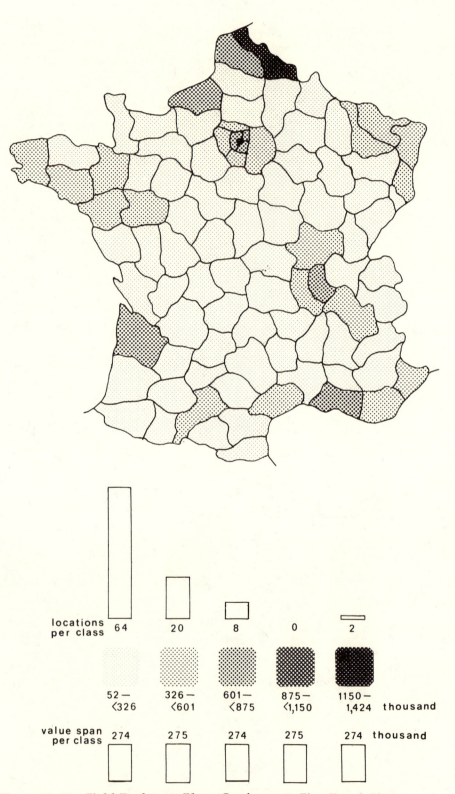

Figure 11–37   Field-Darkness: Plane Conformant, Five Equal Classes

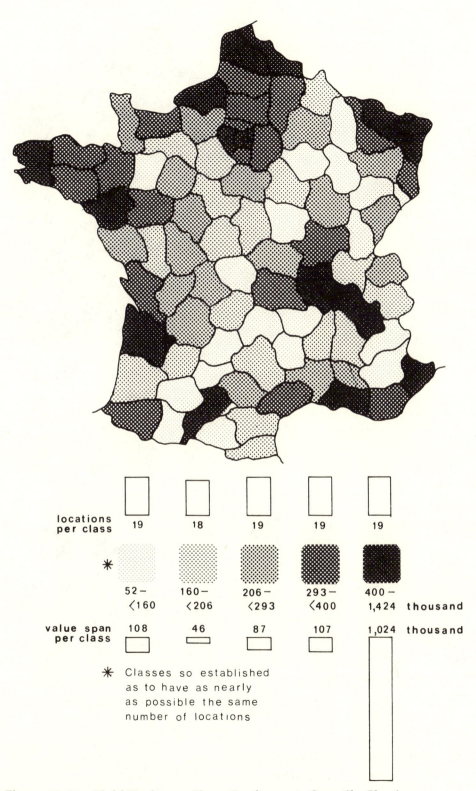

Figure 11–38   Field-Darkness: Plane Conformant, Quantile Classing

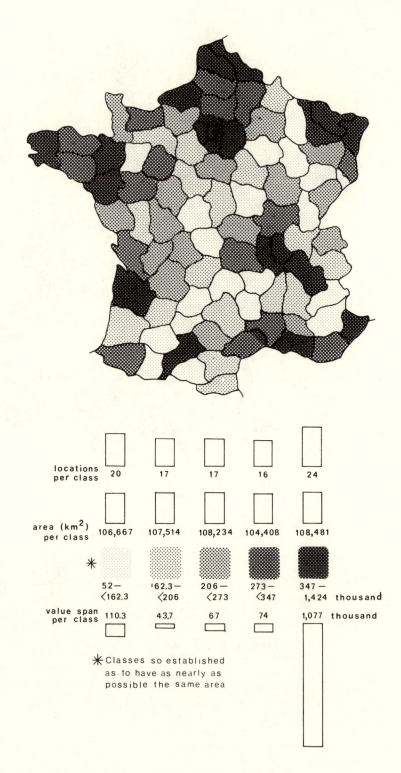

Figure 11–39   Field-Darkness: Plane Conformant, Classed by Equal Area

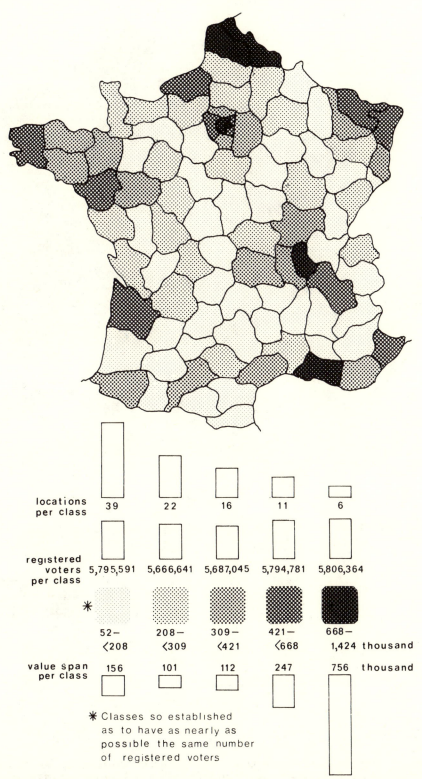

Figure 11–40 Field-Darkness: Plane Conformant, Classed by Equal Number of Voters per Class

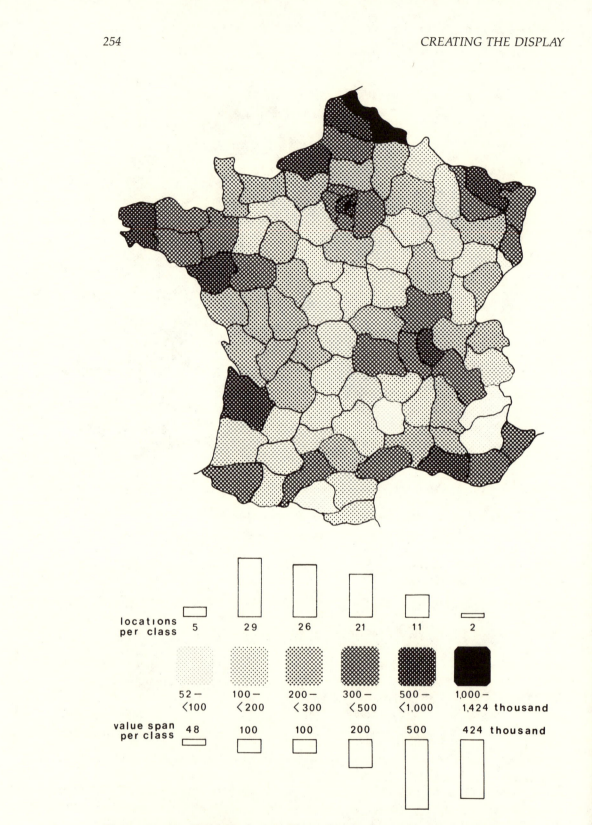

Figure 11–41   Field-Darkness: Plane Conformant, Six Unequal Classes;
Rounded Class Intervals Provide Easily Remembered Class Breaks, in Lieu
of Equal or Quantile Classing

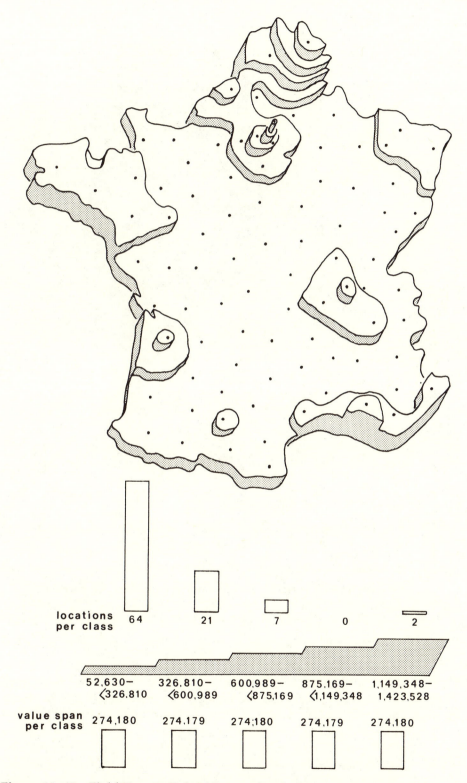

locations per class: 64, 21, 7, 0, 2

value span per class:

52,630– <326,810	326,810– <600,989	600,989– <875,169	875,169– <1,149,348	1,149,348– 1,423,528
274,180	274,179	274,180	274,179	274,180

Figure 11–42   Field-Extent: Raised Interpolated, Stepped, Five Equal Classes (Note Lack of Differentiation Over Much of the Map)

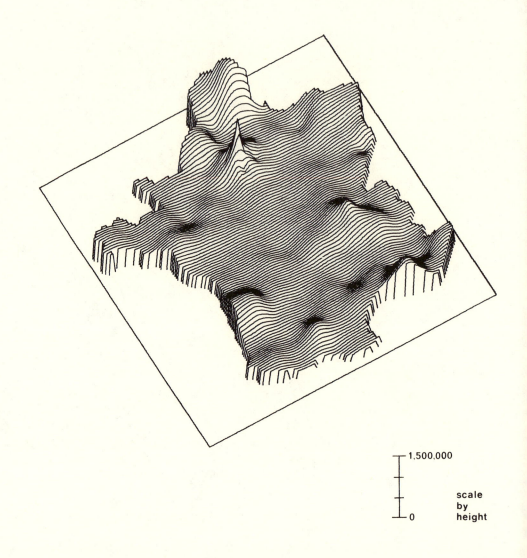

Figure 11-43   Field-Extent: Raised Interpolated, Unclassed

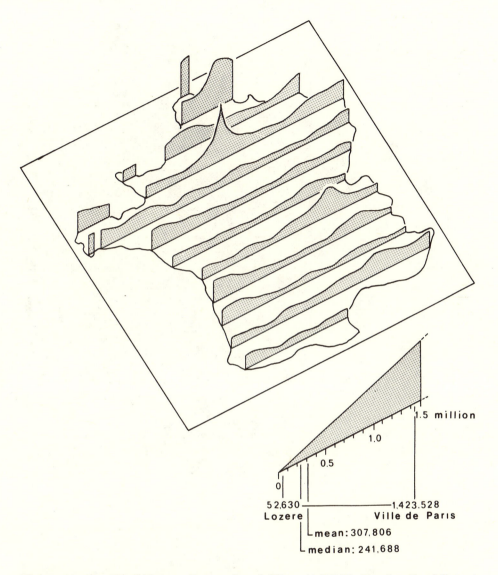

Figure 11–44   Band-Extent: Interpolated Slices (4-BE12). Twelve Equidistant Cross Sections from Figure 11–43

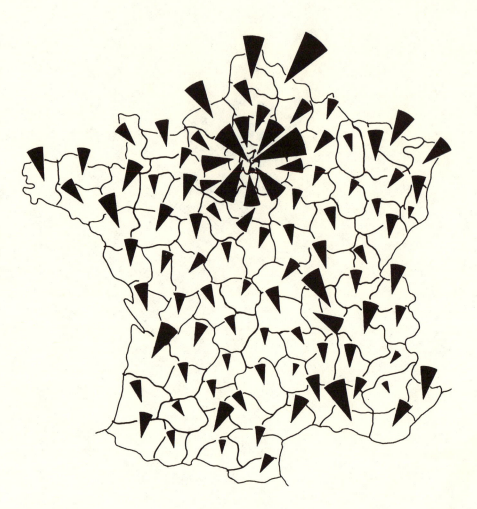

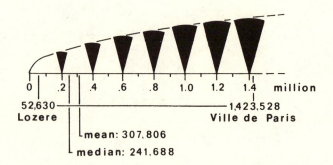

Figure 11–45   Spot-Extent: Sectors, Judged by Area. Circles and Bars Would Be Difficult to Position without Crowding or Overlap

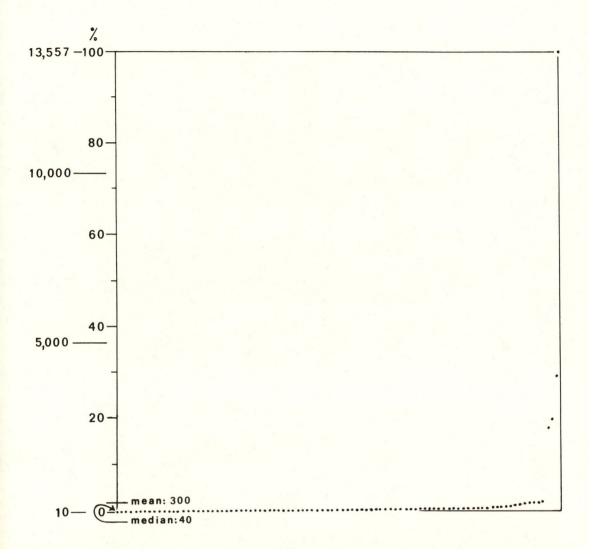

Figure 11–46   The All France II Value Curve: Registered Voters per Square Kilometer

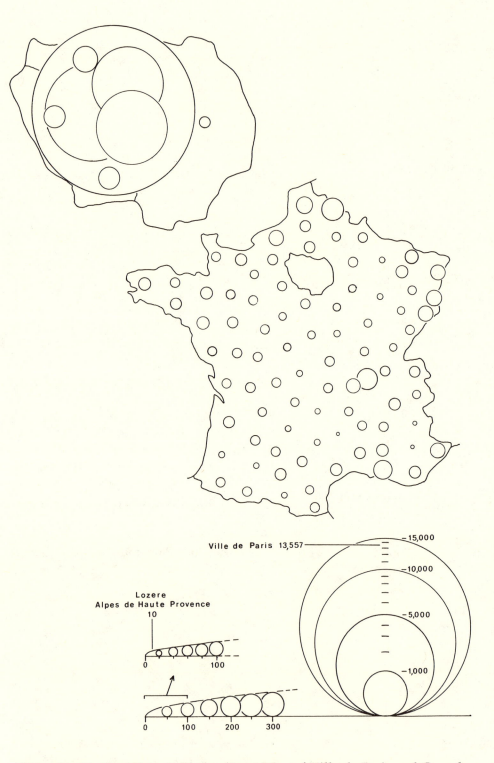

Figure 11–47   Spot-Extent: Circles (Inset Map of Ville de Paris and Complex Value Key are Required)

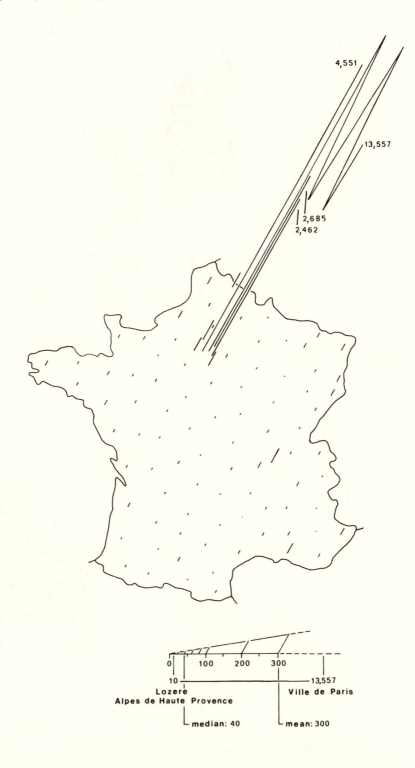

Figure 11–48   Spot-Extent: Bars (Highest Values Are Labeled Because They Cannot Be Measured from Key)

Figure 11–49   Spot-Extent: Sectors (Two-Part Value Key is Required to Measure High Values)

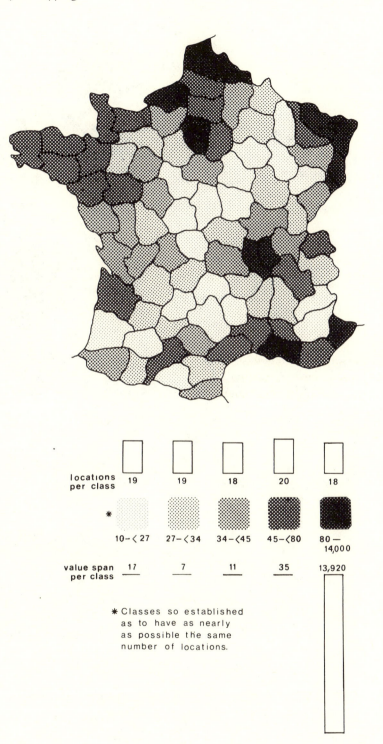

Figure 11–50A   Field-Darkness: Plane Conformant, Quantile Classing

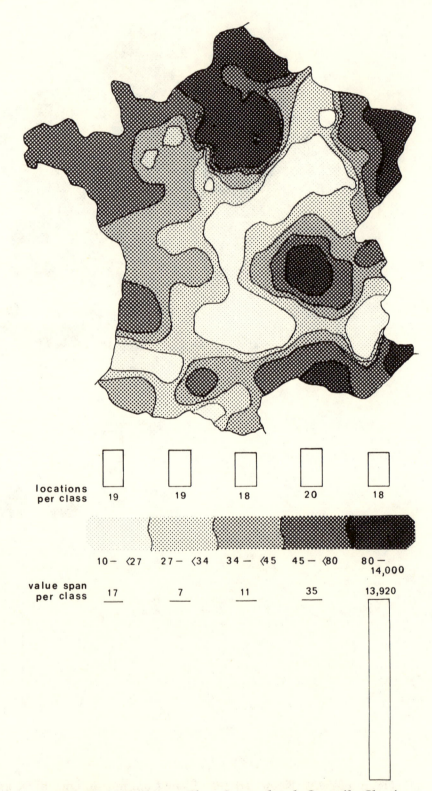

Figure 11–50B   Field-Darkness: Plane Interpolated, Quantile Classing

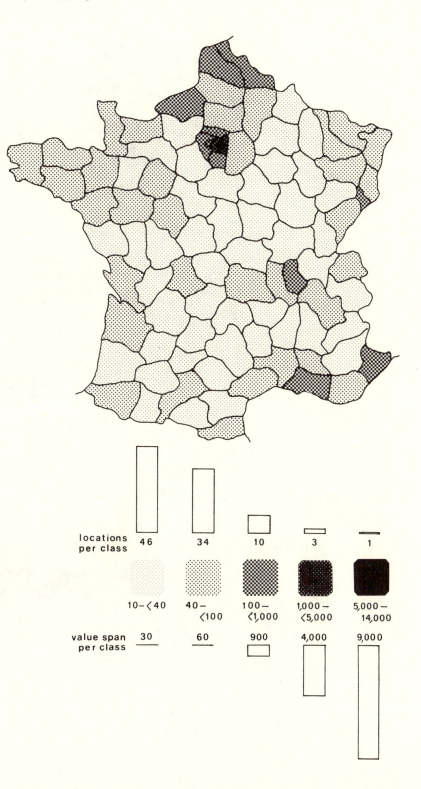

Figure 11–51A   Field-Darkness: Plane Conformant, Rounded Class Intervals

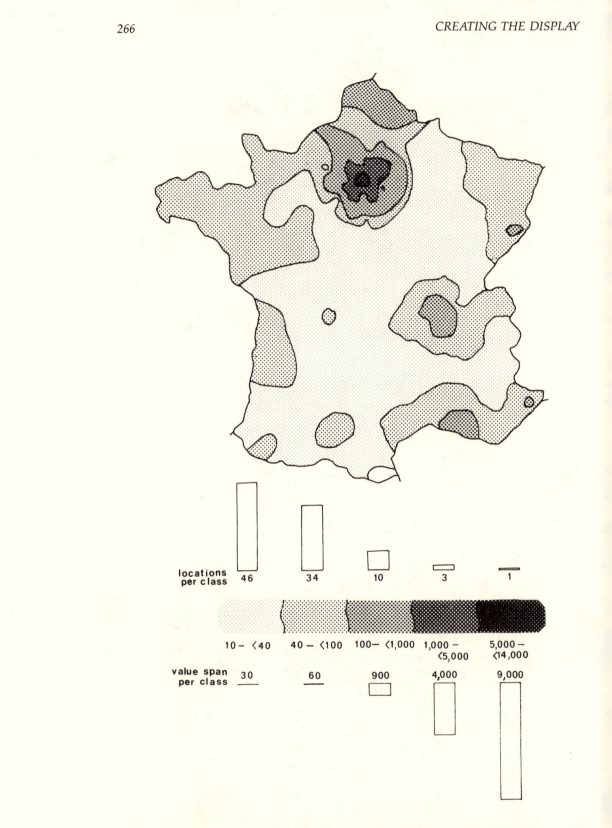

locations
per class   46   34   10   3   1

10 — <40   40 — <100   100— <1,000   1,000 —   5,000 —
                                        <5,000   <14,000

value span   30   60   900   4,000   9,000
per class   _____   _____

Figure 11–51B   Field Darkness: Plane Interpolated, Rounded Class Intervals

CHAPTER 12
Multi-Subject Mapping

A NOTE ON PROCEDURE

For multi-subject mapping, refer to the procedural outline in Chapter 11, Single-Subject Mapping. Handle each specific subject as though for a single-subject map, except as follows:

1. If the study space varies among the subjects, establish an overall study space large enough to embrace all subjects.
2. To the extent that locations may vary among the subjects, assign separate locations.
3. Choose value symbolisms for each subject which are easily differentiable, with those for any given subject of consistent character.
   A. If tone symbolism is used, be sure that tones are not mutually exclusive; each tone must be separately distinguishable even if tones are overlaid.
   B. If spot symbolism is used for more than one subject, choose different, easily distinguishable shapes or colors for each.
4. Maintain the same relative positions among subjects consistently, placing symbols as near base centers as practical.

APPROPRIATENESS AND LIMITATIONS

Multi-subject mapping is desirable in very few circumstances, because of the difficulty of comprehending subjects simultaneously. It is used when the relationship of the subjects must be shown, e.g., land in agricultural use and land used for summer wheat farming in the Dakotas, or land for livestock raising and land for feed lot operations in Boulder, Colorado. Even when showing a relationship is important, however, there is usually a limit beyond which the difficulty of comprehending two or more subjects exceeds the value of being able to relate them simultaneously.

The more subjects to be dealt with simultaneously, the less the comprehensibility of the map as a whole or of substantial portions of the map. A map of monthly temperatures for the United States by states, for example, would allow little overall comprehension. It would be impossible to look at the map as a whole and be able to compare the temperatures in September (or even November) to December. It would be possible, however, to look at any given state, or two or three adjacent states, and read the temperature variability throughout the year. A similar map dealing only with the four seasons (in contrast to the twelve months) would present fewer problems, but for most purposes, it would be even more useful to show a single-subject map of the mean annual temperature for each state.

Multi-subject mapping is warranted under only one other circumstance, namely, when emphasis is on specific data—for each individual location or for a few locations close together—and general comprehension is not required. Under such circumstances, the limitations of the method can be overlooked and a number of subjects can be shown simultaneously.

Each type of symbolism has its own advantages and disadvantages. In general, color will prove more advantageous with multi-subject than with single-subject mapping, as in the use of a different color for each subject being mapped (see Part IV, Practical Aesthetics).

Without color, differentiation among the subjects will have to depend on the positioning of the symbolism as well as variations in shape, tone or texture, shades of gray, or some combination of these (Figures 12–1, 12–2, 12–3, and 12–4).

In using SPOT-EXTENT symbolism, it is important to always show the different subjects in the same sequence. For example, if temperatures per month are to be represented, always begin with January and continue progressively through December, working from left to right. In an example such as this, position alone should be sufficient to differentiate the different subjects.

If the subjects do not have an obvious flow and relationship, fewer subjects can be meaningfully shown. It would probably be unsuccessful to show ten or twelve different socioeconomic factors, for example. A maximum of four or possibly five subjects might be practical. Position alone would probably be the basis for differentiation among the different subjects, and no advantage would be gained by using different shades of gray or different textures or coarsenesses.

With SPOT-DARKNESS, it is usually not possible to use tone to differentiate between subjects if tone also has a quantitative function. Differences in coarseness or grain might be used, but in general these are not promising.

Band and area symbolisms are rarely used for two or more subjects. For example, in a map combining the subjects of railroad, motor vehicle, and barge traffic, the arterial systems for each separate subject are likely to be superimposed and intertwined. With a majority of area symbolisms, simultaneous display of the subjects over the same area would result in a map which was almost impossible to read.

Let us now consider each separate type of symbolism for multi-subject use in terms of black-and-white mapping.

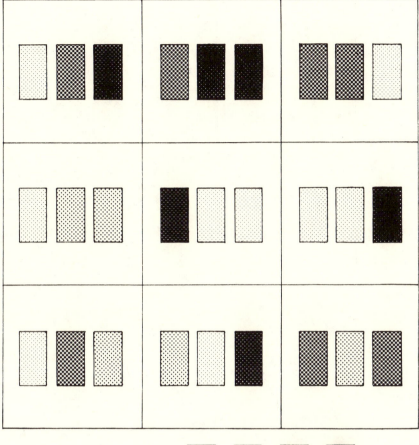

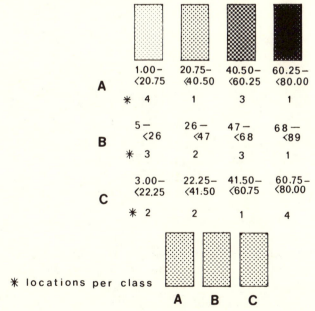

Figure 12–1 Subject Differentiation by Symbol Position

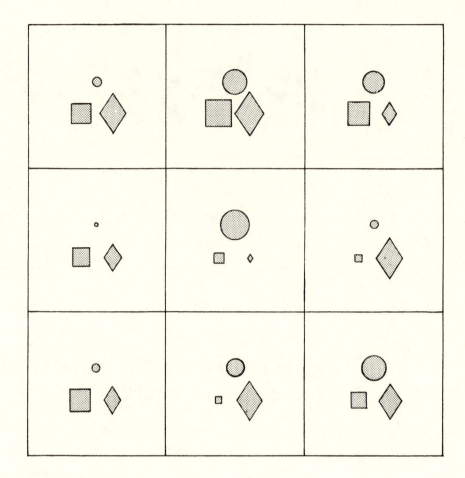

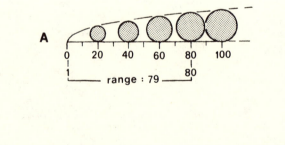

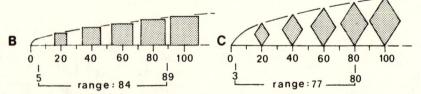

Figure 12–2   Subject Differentiation by Symbol Shape

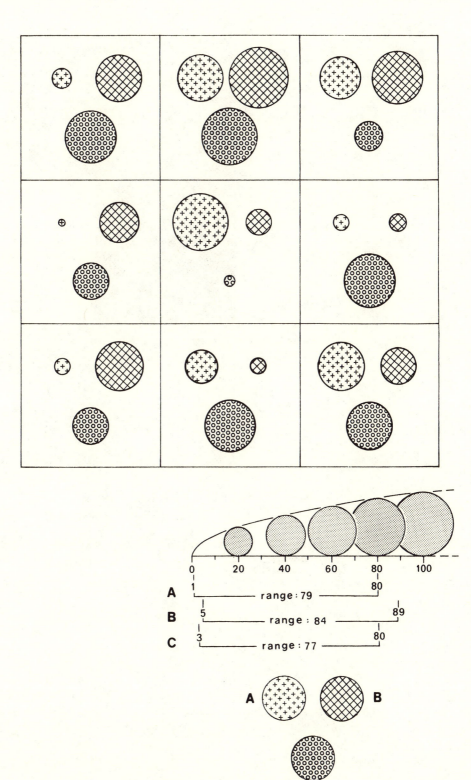

Figure 12–3 Subject Differentiation by Symbol Texture

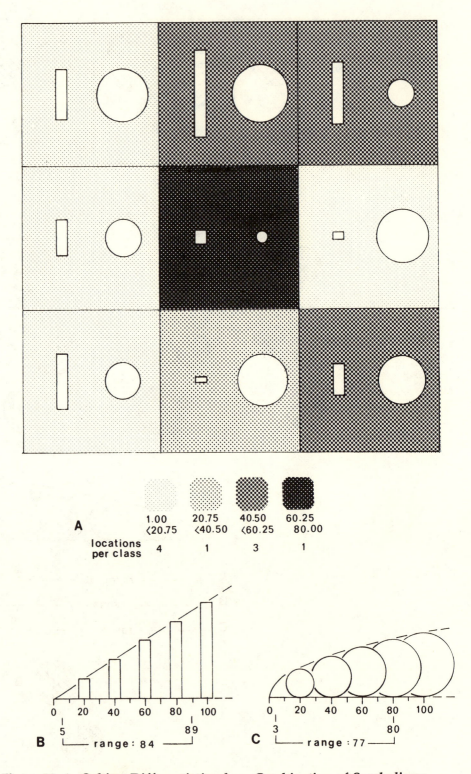

Figure 12–4   Subject Differentiation by a Combination of Symbolisms

SYMBOLISMS

Spot-extent: Bars

Because two or more bars can be placed adjacent to one another—closely spaced or actually touching—bar symbolism is one of the most useful types of multi-subject symbolism when appropriate (Figure 12–5). The use of up to ten or twelve bars usually allows satisfactory comprehension, provided the subject matter has a logical and obvious relationship, such as the example in the previous section of temperatures by month. To add clarity in that example, it might be desirable to group the bars into four clusters of three each—each cluster representing one season. The map user could then find any given month more easily, and some comparison between seasons might also be possible through the visual impression of each separate cluster.

The main advantage of bars is that they vary in length only, as compared to other EXTENT symbolisms which vary in area or volume, so that differences among subjects are clearly visible. The main disadvantage of bars in multi-subject mapping is that they lack personality. Remembering what each bar stands for is difficult except when there is a logical relationship, as in the example of monthly temperatures. There may also be some loss of clarity if class intervals are to be shown, since the more detail provided for any one bar, the greater the competition for overall comprehension.

Squares and Circles

While bars symbolizing different subjects can be placed close together and compared directly, the relative sizes of squares and circles are harder to judge (Figure 12–2). As a result, squares and circles are usually of little value for multi-subject mapping.

Sectors, Fixed Radius

Sectors with fixed radius may also be used to juxtapose different subjects. Direct comparison of both the arc and the area for each subject is easy, and the location to which the symbolism applies is particularly obvious because the common points of the sectors coincide with the base center of the location.

This type of symbolism is particularly useful when the various subjects can be aggregated to 100 percent. Thus, in every instance a complete circle could represent 100 percent, with the sector subdivisions of that circle representing both the absolute quantity of each subject and the relative proportion of each subject to the whole. This is the traditional "pie" type of diagram which is with justice used extensively. Relative position among the subjects should be maintained. If tones of different darkness are employed, those subjects represented by the darker tones will in general receive greater emphasis. In special cases this may create no problem or even be advantageous. In general, however, tones of different coarseness will be a better choice than tones of different darkness. Color may prove particularly helpful here, as will be discussed in Chapter 14.

The choice between bars and sectors may depend primarily upon whether the mapmaker wishes to emphasize the total aspect with sectors or the comparative quantity for each subject with bars.

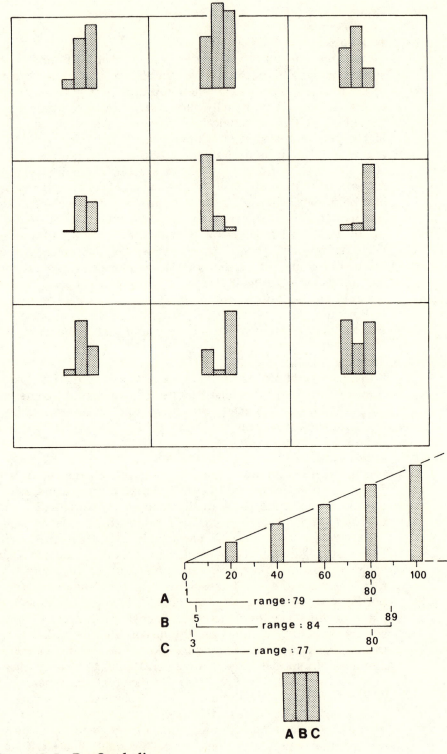

Figure 12–5   Bar Symbolism

Sectors, Fixed Arc

This symbolism combines some of the advantages of bars with some of the advantages of sectors, and has the special virtue of applying to situations where two or more data zones lie extremely close together. Replacing the bars or circles adjacent to one another for each of two or more data locations would involve departure from true position that would seriously affect the readability of the map in terms of location. When circles and squares will not fit, a fixed-arc sector can touch a location with its pointed end. Thus, while the symbol may not be centered on the location, it usually points to it. This would be helpful, for example, in mapping attendance in different divisions of educational institutions, which are often clustered in cities. In very tight situations, the designer should choose the arc accordingly. More acute sectors will allow greater flexibility in placement. Sectors can also be rotated to avoid overlapping symbols. In any case, the different quantities for the subjects are shown by varying the area of the sector only. The arcs must remain constant, so as to avoid the confusion which would result from two simultaneous variables. Here again the same relative position should be maintained within each cluster of subjects per data location. Differences in darkness or coarseness of tone may not be necessary.

Shapes, Miscellaneous—Depictive

When the subject matter lends itself to representational shapes and the special drafting involved seems warranted, this symbolism is an excellent choice. If one were mapping the relative quantities of different types of domestic animals, for example, it might be possible to depict cows, sheep, and goats, varying the size of each while still preserving relative position throughout. This approach is not usually effective with such subjects as the percentage of the population in different ethnic groups or changes over time. It is important to note that this approach normally depends on 2-dimensional variability, namely area. Adjusting the size of the symbol according to height only, for example, is likely to mislead. (A symbol of a human figure that is twice as high as another looks more than twice as large.)

Cubes, Spheres, and Cones

Cubes are usually inappropriate for multi-subject mapping (except perhaps with color). Position alone is likely to be effective, thus limiting use in most cases to only two or three subjects. As 3-dimensional symbols can usually benefit from shading, using different tones or grains to differentiate among the symbolisms may interfere with their depiction.

The problem with spheres and cones is similar to that just discussed, except that shading is even more important. These symbols are not effective for multi-subject mapping.

Extent: Band

As was already noted, line symbolism applied to linear values such as information about transportation routes is likely to be interwoven. Therefore, the problem of multi-subject mapping becomes difficult. Color will usually help, as would differences of tone or grain, but subjects represented by darker tones tend to assume greater importance in the eye of the user. This may be advantageous when different emphases are desired, but as differences in darkness suggest differences in quantity, and black lines appear wider than gray lines of the same width, a problem arises.

Slight changes in shape along the edge of the lines might be legitimate. For example, traffic on railroads might be shown with a straight or continuously smooth side margin, and traffic on rivers given a slightly undulating edge suggestive of water. Traffic on railroad lines might be shown with a slightly serrated edge suggesting railroad ties. A dashed line might be used in contrast to a solid or continuous line—or long dashes in contrast to short dashes. There is a limit to how far this approach can be carried, however.

If the subjects are the traffic patterns in opposite directions rather than different kinds of traffic, two subjects could be shown by varying the width of the bands on each side of the centerline down the middle of the traffic artery. The centerline could be white and the bands on each side a uniform shade of gray—or with maps of relatively small grain, black.

Extent: Line

In contrast to the previous symbolism, height is used instead of width, and the problem for multi-subject mapping is still more difficult. For two or three subjects, bands might be placed on top of one another, distinguished by position and tone. Grain would probably be confusing and therefore undesirable. This symbolism is too complex for more than two or three subjects.

Interpolated Symbolism

Two subjects might be mapped simultaneously, using lines of narrow width for one and lines of a slightly wider width for the other, or long dashed lines for one and continuous lines for the other (though this is less likely). Color can be used effectively in this type of mapping.

Although this symbolism is extremely readable and clear for single-subject mapping, the simultaneous depiction of two different 3-dimensional shapes somewhat resembles a double exposure on a single negative. Greater comprehensibility would result if the lines of intersection between the two surfaces were shown. However, in most cases this would involve considerable difficulty in drafting. Under rare circumstances, the result might be fully warranted and especially effective, something like a phantom view. The highest portions of the surfaces for each of two subjects could be shown in solid lines and the hidden portions shown in very thin lines, which would contribute greatly to the clarity of the map.

Dots

It is possible to use dots of different shape, each shape associated with a different subject. For example, small circles might be used for mapping one kind of agricultural product, small squares might be used for mapping another, and small triangles for a third. Or hollow dots might be used for one subject and solid dots for another (though this would definitely emphasize the subject represented by the solid dots). However, more than one shape or color of dots is likely to be confusing because of the difficulty in focusing attention on any one subject. Unless the primary interest is the relative spatial density among subjects, it will be preferable to make a different map for each subject.

Spot-extent Combinations

A special opportunity may arise in multi-subject mapping for using sectors that vary in radius and arc at the same time. For example, a change of arc might represent a change in the quantity of a population in different income levels, and a change of radius might represent a change in income levels. Thus the farther any given sector projects from the center, the higher the income level represented by that segment of the population, and the wider the sector (i.e., the greater the arc), the greater the population possessing that income. This type of symbolism must, of course, be used with classed data. Either arc or radius should be related to area, but only one need be classed, preferably radius. (Note that bars could likewise be varied in two dimensions, width and height—width representing one variable and height the other.)

A similar procedure could be followed with three quantitative variables in terms of other shapes. For example, 2-dimensional bars could be varied in thickness to represent a third quantitative variable. Whenever three variables are involved, a 3-dimension type of representation can be used— isometric or perspective. With two variables only, a cylinder of varying size and varying thickness might be used. The diameter of the circle might represent the population and the height of the cylinder might represent the mean income level of that population. With all of these SPOT-EXTENT combinations, overall comprehension will be difficult, and therefore these approaches should be limited to maps where emphasis is on data applicable to individual locations.

PART IV

Practical Aesthetics

CHAPTER 13

Designing in Black-and-White

GRAY SCALES AND DOT SCREENS

Editor's note: David Collins, mathematician, and Herbert C. Heidt and Eliza Mc-Clennen, cartographers, assisted in the preparation of this section.

The single most effective quantitative analogue employed in mapping work is darkness; the darker the tone, the greater the quantity represented. However, pure grays without visible pattern are of very limited usefulness, since most people cannot differentiate among more than four or possibly five tones at most, especially if the tones are randomly placed, as in conformant mapping. A visible pattern composed of black and white (without gray) is essential to assist in differentiation. However, to be successful and aesthetically acceptable each pattern must not be too coarse. Assuming pattern elements to be of one constant shape, questions arise as to:

1. The best shape and arrangement of shape to employ.
2. The optimal degree of coarseness so that the pattern is not too fine to be effective nor too coarse to be acceptable.
3. The percentage of ink coverage to employ for each of the patterns.

Our research suggests the following answers:

1. Shape of pattern elements: the best single shape without doubt is the circular dot—black on white, perhaps combined with white on black. As to disposition of elements, a square pattern is superior to a triangular one.
2. Degree of coarseness: dots at approximately 32 to 42 per linear inch appear to offer the best compromise. If coarser, the result (though still legible) becomes aesthetically unacceptable for maps. If finer, the benefits of pattern are lost (Figure 13–1, left to right).
3. Ink coverage: when more than three or four patterns are to be employed,

*See also Chapter 15, Reproduction Processes.

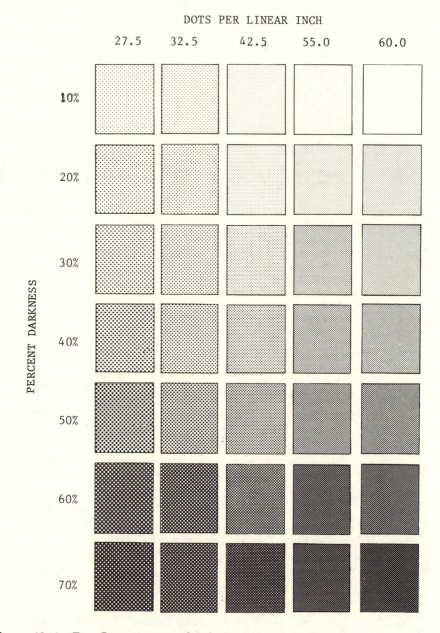

Figure 13–1   Dot Coarseness and Ink Coverages

the smallest dots capable of being well reproduced should be used for the lightest and darkest patterns (black on white and white on black), in order to provide maximum differentiability among the successive patterns (Figure 13–1, top to bottom).

However, serious problems remain as to the percentages of ink coverage to employ for the intervening patterns. The obvious solution, spacing them equally between the extremes, gives a visual impression of substantial irregularity. The next most obvious solution, increasing ink coverage step by step according to fixed ratios, is equally unsatisfactory.

The achievement of equal visual spacing with dot patterns has been the subject of considerable study among geographers and cartographers, but no firm conclusions have been generally accepted. The most widely, though by no means universally, accepted solution appears to be that developed at Harvard some years ago by Robert L. Williams, then a Ph.D. student in geography. His solution, based on empirical research, proves to be very close to our theoretical solution. Further study suggests inherent conformity of the two solutions.

The heavier solid line of Figure 13–2 shows the theoretical (and inversely symmetrical) gray-scale curve developed pursuant to our research, and the ligher dashed line shows the trend of the empirical data resulting from Williams's researches. The straight-line segment curve shows the gray-scale

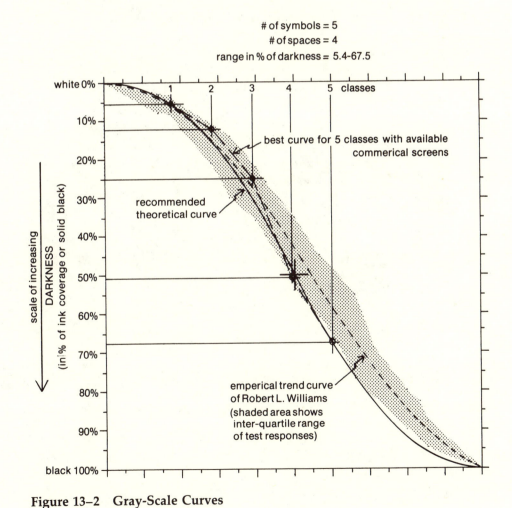

Figure 13–2 Gray-Scale Curves

curve of commercially available screens in contrast to the theoretical curve. Use of this last-mentioned curve allows visually equal spacing to be achieved between any number of gray-scale tones, corresponding to the number of classes desired. Figure 13–3 shows a map in five classes using commercially available stick-on dot patterns that were selected to approximate the ideals suggested by the curve. The patterns appear approximately equal visually and sufficiently differentiated to permit easy distinguishability. (With a large number of classes, the latter point becomes crucial.)

ELEMENTS OF EFFECTIVE SYMBOLISM

Those symbolisms likely to be most helpful in either actual use or from a conceptual viewpoint (in establishing a framework for thinking about symbolism) must be capable of being interpreted over the entire map. A symbolism is poor if each element or small cluster of elements must be studied individually. Since conformant symbolism requires that each area be judged

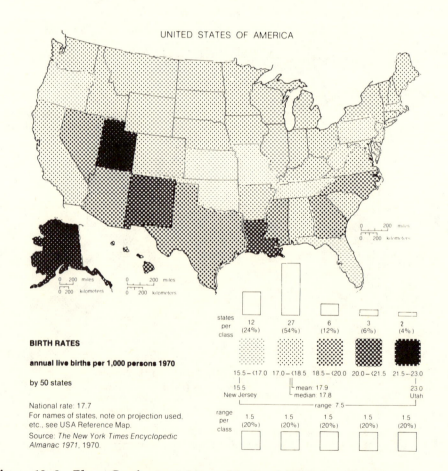

Figure 13–3 Plane Conformant Map with Good Darkness Variation

independently, we will use it as the test case in the following discussion.

Certain prerequisites are unavoidable. Classing is essential to achieve easily differentiable FIELD symbolism, as a full range of values would be too hard to judge. FIELD symbolism should be relatively fine-grained—fine enough so that the final display can be modestly sized, and fine enough to fit within the boundaries of small geographic zones.

The following elements create the tones of FIELD symbolism when combined effectively: darkness, size (EXTENT), centering of symbol elements (coarseness), shape, disposition (variable orientation, square vs. triangular, straight vs. curved, positive vs. negative), and countability. None of these is adequate alone, but successful combination avoids the pitfall of meaningless elements.

Darkness alone (greater darkness, greater value) gives inadequate differentiation for more than three or four classes. No matter how many classes, darkness must be easily differentiable and at least roughly proportional to the quantity represented. Although it is the most powerful basic variable for FIELD symbolism, pure darkness should be reinforced or supplemented in one or more ways.

FIELD-EXTENT (greater element size per pattern, greater value) cannot practically be used alone, as variation in darkness will always be produced unless the coverage is kept constant. It is, however, an invaluable reinforcement for darkness.

Varying the shapes of the elements forming the patterns can prove extremely useful in differentiating among patterns of grays. For example, the gray tones of computer maps produced on line printers have of necessity used available numbers, letters, punctuation marks, and so on, alone or in combination. The resulting grays have seldom proved entirely successful, but varying the shapes of the elements has been the most important factor in the degree of success that has been achieved.

Combining elements of variable, distinctive shape with associated percentages of ink coverage should make it possible to use ten classes successfully—in contrast to a maximum of five or six classes when uniformity of shape is maintained. Figure 13–4 illustrates a hypothetical map with ten classes based on distinct shapes. The first group of five is positive (black on white), and the second group of five reverses the same pattern to negative (white on black).

Field disposition (variable orientation) must be easily differentiable when used alone, but since it is not obviously quantitative, it is somewhat distracting. It is not useful with more than four classes, but it could be used as a supplement to darkness (Figure 13–5).

The positive-negative element is not useful for FIELD-EXTENT symbolism by itself, as there are only two possibilities, and it also increases the difficulty of showing zone boundaries, lettering, and so on when used alone. It could be used as a supplement to darkness to differentiate symbolism above and below the midpoint with an even number of classes. With an odd number of classes, the middle class could be "checkerboarded" if the positive and negative ink coverage are reciprocal (Figure 13–6).

FIELD-COUNT is very useful in combination with darkness. In this special

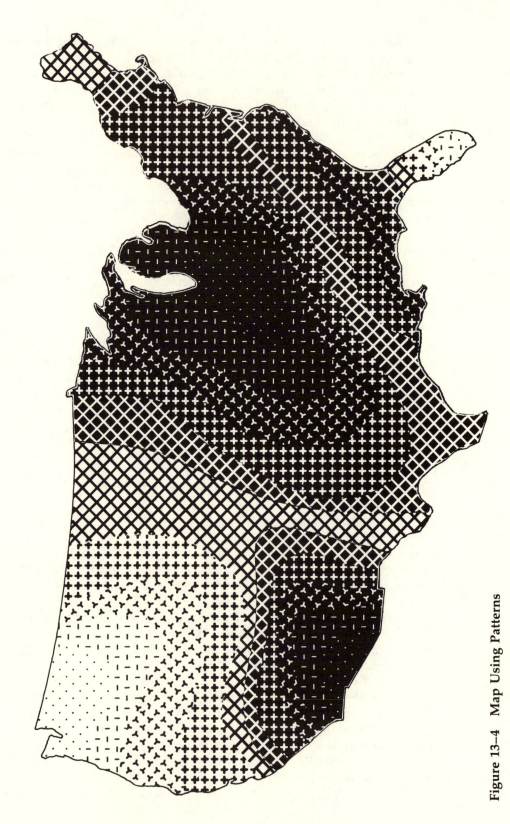

Figure 13–4 Map Using Patterns

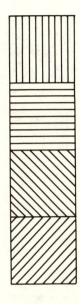

Figure 13–5A   Field-Disposition, Variable Orientation

Figure 13–5B   Field-Disposition, in Combination with Darkness

set of symbols developed by the author, not only is the shape varied to aid differentiation, but the number of points per element has been designed to conform to the number of the class being represented (Figure 13–7). The tone employing elements with four points, for example, represents the fourth class. By this means, the analogue of variable darkness is achieved when

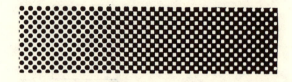

Figure 13–6   Positive-Negative Continuum through a Checkerboard Middle

Patent #4,148,507   April, 1979

Note: The system of symbolism will be available in the form of self-adhering
      preprinted sheets from Letraset Limited and its worldwide subsidiaries—
      in the United States, Letraset USA, Inc., 40 Eisenhower Drive, Paramus,
      New Jersey 07652.

Figure 13–7   Fisher Symbolism; Patent Number 4,148,507

attention is directed to the map as a whole, yet the class of any given tone may be determined with ease (Figures 13–8 and 13–9).

Variables that are random or arbitrary, and therefore meaningless, must be avoided. The basic objective for black-and-white maps is to produce a series of gray tones that provide maximum differentiability. If differentiability

0–<20 20–<40 40–<60 60–<80 80–100

Figure 13–8 Foursquare Map Using Fisher Symbolism in Five Classes

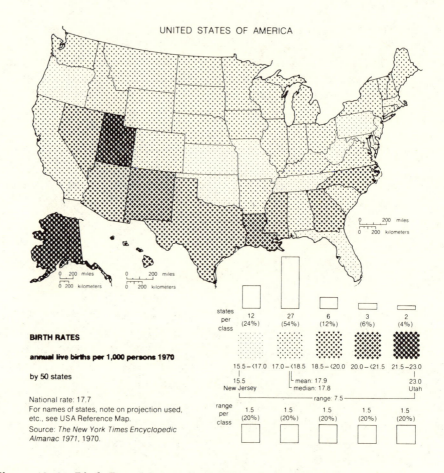

Figure 13–9   Birth Rate Map Using Fisher Symbolism in Five Classes

depends on (1) darkness alone or with change of size and arbitrary shape, it will be inadequate; (2) darkness and change of size alone, it will be better but still inadequate; (3) darkness, change of size, and change of direction, it will be better, but an arbitrary and meaningless variable has been introduced. Darkness with change of size and meaningful shapes, aided by count, will be most successful. When the designer is not limited to black and white, the addition of progressive change in color is optimum.

CHAPTER 14

Color for Quantitative Differentiation

Editor's note: For a more complete discussion of color theory, see "An Introduction to Color" and "The Form of the Color Solid" by Howard Fisher in Volume 6, *Thematic Map Design*, of the Harvard Library of Computer Graphics, 1979 Mapping Collection.

INTRODUCTION TO COLOR

For any single-subject map, variability in quantity may be represented by variability in color. Such variability may be demonstrated in terms of color-hue, color-intensity, or both, usually combined with variability in darkness. Differentiation by color may be effected with TONE symbolism of all three types: SPOT, BAND, and FIELD.

Multi-color is valuable as a quantitative analogue in two principal ways. First, multi-color is capable of providing far greater visual differentiability between varying value levels than could ever be achieved by the use of black (BK)-and-white (W) alone or any single color. Imagine a map for which the values were grouped into three classes represented by three progressively darker colors in spectrum sequence, such as yellow (Y), orange (O), and red (R). Such a map would have a greater graphic impact and could be read more quickly and easily than any comparable map in which gray (GY) or some other single color was employed in three progressively darker tones. Not only would the overall trends be evident, but the distinct hues would allow the viewer to visually extract areas of a common class.

Second, multi-color combined with another quantitative analogue will permit a far greater number of classes to be employed than could ever be achieved by means of gray alone or any other single color. For example, five well-produced colors might be used, each in three clearly differentiated levels of darkness—for a total of fifteen classes.

Before proceeding further to explore the uses of color as a quantitative analogue, we must consider in some detail the ways in which color-hue and color-intensity can be varied systematically. It will be particularly important to explore the interrelationships existing between those variables and darkness.

The field of color has long been noted for the vagueness and inadequate standardization of even its technical terminology. We will examine the precise meanings of several technical terms which will prove useful, some of which we have already employed in a general sense. For our purposes, the four basic terms are *dominance, hue, darkness,* and *intensity.* These will be considered in the order mentioned, but before proceeding a few introductory words are necessary.

Assume that the colors referred to are those most characteristically associated with particular designations. Thus, when we say *red,* we mean a usual red-red not a red-orange (RO), red-purple (RP), or a red of special character. Except where otherwise stated, assume that each color is of high intensity. Further assume, unless stated otherwise, that each color is of that darkness most likely to be associated with its hue at such high intensity. Thus, *red* means a bright and therefore somewhat dark red, not a pink or dull reddish brown or an unusually dark red.

In these examples, there is obviously great latitude for interpretation. No two persons would probably agree on what precise tone is intended by red, but they might be able to agree on the *approximate* tone intended. For many purposes, the mapmaker will wish to define colors as precisely as possible. It is very important, however, to keep in mind that you may not be able to achieve the precise colors you wish and that the map user should not have to be a color expert. The mapmaker needs easily produced colors. The map user needs obvious differences of a clear and striking character. Clear, distinctly enunciated, and easily comprehended graphic statements are required. Delicately modulated and perhaps poorly comprehended subtleties are likely to convey little more meaning than a mumbled telephone conversation over a bad transatlantic connection.

It is generally preferable to avoid colors that are difficult to name, or exotic, or unexpected in context. Common unambiguous color names are not only easier to locate on the map but assist overall comprehension, as well as make the value key simpler and more memorable.

Considerable leeway will always be available in choosing precise hues, darknesses, and intensities. It is hoped that the examples of multi-color maps discussed in this book will be regarded less as models to follow than as possibilities to improve upon.

Dominance

Though not normally considered in color analysis, the factor we will designate *dominance* can at times prove critical when selecting color schemes.

Imagine a variety of relatively high-intensity colors placed more or less at random on a white background. Even at a casual glance, certain colors are far more dominant or attention-getting than others. In general, reds, red-oranges, and oranges stand out particularly, and therefore those colors have traditionally been employed for safety and warning signs. In contrast, greens, blues, purples, and yellows, when viewed directly against a white ground, tend to be inconspicuous or *subdominant.* Yellow-orange and red-purple are typically of intermediate dominance. All colors of relatively low intensity will be subdominant.

Under normal circumstances, in establishing any progression of high-intensity colors to represent increasing values, increasingly dominant colors should represent increasing value. Any series of colors involving high-intensity red should build up to and end with that color. In any color series that includes red, red-orange, or orange and ends with red-purple, all the reddish hues should be grayed or muted (i.e., reduced in intensity) to be less dominant than the terminal red-purple.

Yellow is the starting point for many satisfactory color schemes. It must be carefully chosen, however. A pale, weak yellow will not show up clearly, whereas one of a high intensity may appear too dominant to represent the lowest class.

A graphic representation of the main facts regarding relative dominance may be useful. These two color sequences, each starting with yellow, the lightest of all high-intensity colors, and ending with purple, the darkest of all high-intensity colors, represent the principal range of possibilities. Clearly dominant colors which may cause problems are underlined in capital letters. Less dominant colors are shown in capitals, not underlined.

LIGHTEST	yellow (Y)	yellow (Y)
	YELLOW-ORANGE (YO)	yellow-green (YG)
	<u>ORANGE (O)</u>	green (G)
	<u>RED-ORANGE (RO)</u>	blue-green (BG)
	<u>RED (R)</u>	blue (B)
	RED-PURPLE (RP)	blue-purple (BP)
DARKEST	purple (P)	purple (P)

It will be obvious that black, or near black, is a poor terminal choice for any color series representing quantity. If used, black should ordinarily be employed only with colors of low intensity, so that it is clearly the strongest or most forceful tone. Black would be hardly noticeable in any map using high-intensity hues or medium-intensity red or orange. (This would be especially true if black were also used for text or other purposes.)

HUE

The term *hue* will be used to designate a specific color, such as green, red, or purple. It is the most fundamental aspect of color.

Variation in hue may be represented graphically by the traditional color circle shown in Figure 14–1. The three primary hues are indicated in the diagram by dots with the letters R, Y, and B printed in large type. The three secondary hues (each located on the circumference of the circle halfway between adjacent primary hues), orange, green, and purple, are indicated by dots with letters printed in medium-size type. The six tertiary hues (each located halfway between adjacent primary and secondary hues), red-orange, yellow-orange, yellow-green, blue-green, blue-purple, and red-purple, are indicated by dots with letters printed in small type. The locations of the four colors used in the four-color printing process have been shown by small dots. The printing color near yellow is *process yellow* (PY), that near red-purple is *process magenta* (PM), that near blue is *process cyan* (PC), and that in the center is black.

When selecting a color series to represent a quantitative series, it is usually desirable to follow the hue sequence of the color circle, proceeding in either a clockwise or counter-clockwise direction. This recommendation is made partly because of knowledge regarding spectrum and color circle sequences, but mainly because any such sequence will provide a sense of continuity. Any series of colors so selected may also display a continuously progressive change in darkness. In the diagram the twelve basic hues of high intensity have been spaced equally around the perimeter of the color circle. Yellow, the lightest, is located in the top central position.

It should be stressed that, considering the problems typically encountered

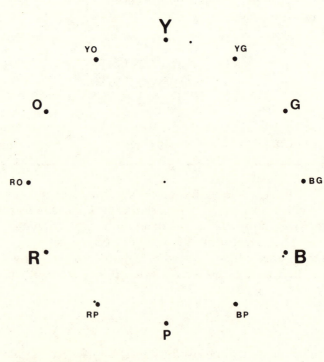

Figure 14–1   Color Circle Diagram

in reproducing colors to desired specification, even the six tertiary hues (those represented by hyphenated designations) should be used with caution. When tertiary colors are employed, finer visual distinctions are required. The problems resulting from simultaneous contrast are likely to prove more troublesome because of the smaller differences among tertiary hues.

The three primary hues are customarily thought of as independent and unique, and therefore should not be used alone in a series as they provide no sense of continuity. Each intermediate hue falling between adjacent primaries appears to possess some aspect of each of those primaries. In every shade of orange, for example, the sense of both red and yellow is present. There is a logical as well as visual flow through all the successive hues around the color circle. No other sequence of hues provides this sense of continuity, which contributes importantly to both comprehension and memory.

The sequence of hyphenated names of the tertiary hues is sometimes the reverse of that here employed. For our purposes, placing the name of the primary color first in each composite designation is easier to remember and avoids favoring a particular directional tendency when thinking about the color circle. In the widely used Munsell color system, orange is called *yellow-red* (YR) or *5YR* (or *15*). Red-orange and yellow-orange have no names and must be referred to respectively as *10R* (or *10*) and *10YR* (or *20*).

DARKNESS

The term *darkness* designates the degree of departure of a color from white. Darkness is equal in importance to hue itself for our purposes, in that hue without simultaneous progressive darkness is often ineffective or misleading. Progressive darkness can serve as a mnemonic device, without which a series of hues may appear arbitrary and be difficult to remember.

Variation in darkness may be represented graphically by a single vertical line (as in Figure 14–2). Darkness is considered *nonexistent* (0 percent) for white paper at the top end of the scale and *total* (100 percent) for black ink or other pigment at the bottom end of the scale. The range between these extremes can best be divided, for our immediate needs at least, into the eight equal spaces shown. These are established by the system of repetitive subdivision previously employed for hue. What darkness percentages result from subdivision into eight visually equal spaces? The problem with multicolor is quite different from the problem with single-color.

As was earlier discussed, with TONE symbolism in gray or any one color, darkness will seldom prove adequate for differentiation by itself, and supplemental devices must be sought. When more than three or four classes are involved, it will usually be desirable to employ as many of the following as possible: visible grain, with elements of variable size on uniform centering; distinctive variation in the shape of the elements; and variation in the shape of the elements, which provides countability.

In establishing our monochromatic darkness curve we were strongly influenced by the consideration of how to increase grain size so that the steps appeared approximately equal. In contrast, color will be visually homoge-

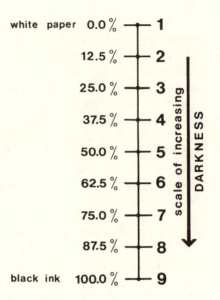

Figure 14–2   Darkness Scale

neous in character and there will be no visible grain. Thus the principal factor which determined the nature of the recommended monochromatic darkness curve is missing when multi-color is involved.

Judging the relative darkness of two samples closely spaced on the color circle—such as blue and blue-green—presents no great problem. However, most persons will often find it difficult to judge the relative darkness of two hues distantly spaced on the color circle—such as red and blue, or orange and green. Since it will ordinarily be most useful to employ colors that are not closely spaced on the color circle, we need be concerned primarily with the achievement of obviously distinguishable steps for use with distantly spaced colors.

In a sense, color is so powerful that the problem of trying to identify small differences of darkness among colors may be likened to the problem of trying to identify small differences of pitch among the voices of comrades in the midst of a cavalry charge accompanied by war whoops and the firing of cannon. Even fairly substantial differences in darkness may not be noticed in the midst of far more conspicuous differences in hue and perhaps intensity. The great advantage of color is that it *is* so much more effective than darkness, which explains why darkness is often relegated to a supportive role not requiring subtle distinctions.

While future research may yield more refined theories on the matter, the soundest course at present appears to be a darkness curve (Figure 14–3) that is a straight line. It will assure equal darkness steps and is somewhat simpler to use than any other possible curve. We recommend it as a guide when using multi-color, realizing that under some circumstances flexibility in application may be appropriate. Such instances pertain to the fact that in the real world, the darkness progression of high-intensity colors from yellow to purple down either side of the color circle is not symmetrical.

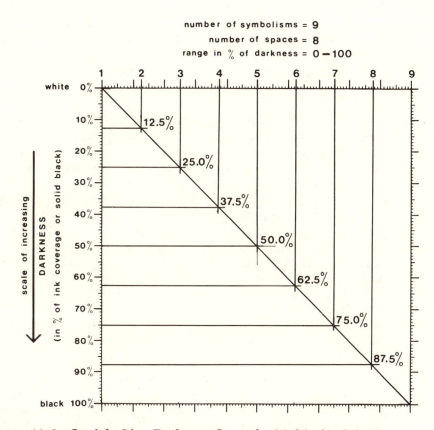

Figure 14–3   Straight-Line Darkness Curve for Multicolor Selection

As a practical matter, it is best to avoid colors less dark than 10 percent or darker than 90 percent. White to black is as much the limit of darkness for each positive hue as it is the limit for gray. Hues close to but reasonably distinguishable from white or black will prove entirely satisfactory.

It is risky to use white to represent a quantity or class, as *quantitative white* and *background white* may be confused.

The only exception might be the representation of missing data or, when there is no possibility of confusion, of precisely zero. Solid black is usually unsatisfactory as a quantitative analogue, as was mentioned earlier.

In general terminology, it is more usual to speak of *light reflectance*, sometimes called *luminous reflectance* but usually simply *reflectance* (or *R*), than of *darkness*. In measuring reflectance, absolute or theoretical black (somewhat blacker than black ink) is used as the starting point (0 percent), with absolute or theoretical white (considerably lighter than white paper) as the terminal point (100 percent). Average black ink and average white paper are 3.1 percent and 84.2 percent, respectively. Reflectance percentages may be converted into our darkness percentages simply by subtracting the reflectance percent from 100 percent and finding the appropriate range. The reflectance figure of 84.2 percent would equal a darkness of 0 percent; that of 3.1 percent would equal a darkness of 100 percent.

In the Munsell color system the word *value* is employed for darkness. This seems unfortunate in view of the inevitable confusion which arises concerning references to other types of value—such as values to be mapped or intensity values. In the Munsell system, ten levels of value are applied to the range between theoretical black and magnesium oxide, the whitest substance known. Munsell darkness level zero (written 0/) has a reflectance of 0 percent and conforms to theoretical black; Munsell darkness level ten (written 10/) has a reflectance of 100 percent and conforms to magnesium oxide. Intermediate Munsell values are unequally spaced in terms of percentage units. They supposedly yield visually equal intervals, but the curve is not logarithmic as is sometimes implied (even in Munsell publications). In the Munsell color system, darkness level 2/ has a reflectance of 3.1 percent and darkness level 9.25/ has a reflectance of 84.2 percent. Since these reflectance values correspond to figures for typical black ink and white paper respectively, and since samples of the darknesses referred to may be procured from the Munsell Company, we have adopted these percentages as the basis for interrelationships that may be useful with the Munsell system.

INTENSITY

The term *intensity* designates the degree of brightness or purity of a hue in the sense of freedom from gray. The use of intensity ratings will alleviate problems due to unsystematic variations in intensity which might prove confusing as well as the use of colors of such low intensity that the usual hue names no longer apply with certainty. Intensity should preferably be constant throughout any color series used as an analogue for quantity, or it should increase with an increase in the quantity being represented. Ideally, such increase should be at a constant rate, but this may be difficult to achieve and it is not essential.

Variation in intensity may be represented graphically by a simple extension of the basic color circle concept. Designate the center of the circle as representing neutral gray. Intermediate levels of intensity then fall between the center and the high intensity hues of the color circle perimeter. Figure 14–4 shows the color circle thus modified. The radiating lines represent the twelve basic hues and the concentric circles represent four levels of intensity.

In the Munsell color system, *chroma* is used instead of *intensity*. It should not be confused with *chromaticity*, which embraces the entire concept of positive color—i.e., hue as well as intensity, in contrast to grays which have neither hue nor intensity. The term *chroma* is also employed as a unit of absolute measurement, and we will use it to refer specifically to Munsell units for the measurement of intensity. Assuming glossy samples, by the Munsell system high-intensity blue has a chroma of eight (/8), while high-intensity red-orange has a chroma of sixteen (/16) (Figures 14–5A and B). These are the approximate extremes of intensity, judged in absolute terms—assuming typical inks or other reasonably permanent pigments.

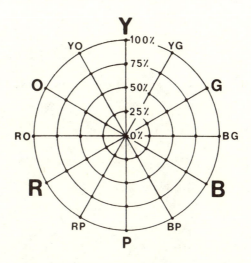

Figure 14–4   Color Circle Diagram with Intensity Levels Added

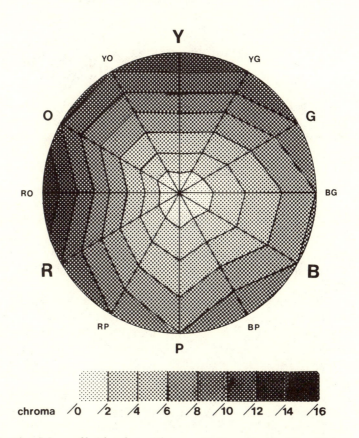

Figure 14–5A   Munsell Absolute Chroma Rating for Twelve Hues

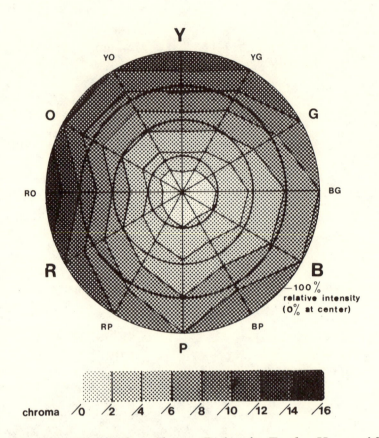

Figure 14–5B   Munsell Absolute Chroma Rating for Twelve Hues, with Relative Intensities Superimposed

It is important to recognize that, in absolute as opposed to visual terms, hues of high intensity are not all of equal intensity. For example, it was just noted that blue at its characteristic high intensity is only approximately half as intense as corresponding red-orange. Visually, however, they are more alike than a red-orange of /8 and a blue of /8, which is the important factor to consider in the design process.

The term *absolute intensity* implies data of known accuracy. This may not be the case. In view of the number and complexity of the variables involved, it is impractical to assign maximum absolute intensity ratings. We prefer to employ the concept of *relative intensity* in order to suggest characteristic circumstances and to avoid any implication of some particular maximum intensity.

Colors of relatively high intensity will typically prove most useful. Such colors are likely to be most easily distinguished from one another. By definition, colors of low intensity tend to be close to gray, and hue differences are greatly subordinated.

SELECTING QUANTITATIVE COLOR SCHEMES

Having established a basis for effective exploration within color space, we can now consider how to establish color schemes.

In seeking the most effective color scheme for any particular project, the four principal variables are likely to be the following:
—The number of colors to be employed;
—The number of classes into which the data is to be divided;
—The type of symbolism to be used;
—The production or reproduction method involved.

In some cases, precise answers to these questions may be established in advance. In that event, the problem is to select the color scheme that conforms most closely to the requirements. Under more typical circumstances, answers may be tentative and color may influence some of the other variables.

Let us consider the first three aspects individually in more detail.

The Number of Colors to Employ

If a map is to be printed, in general, the greater the number of colors, the greater the cost. When maps are run with other color work, however, no economy will be achieved by using fewer colors than are used in the other work. The use of a maximum of four ink colors, either in four separate runs through a small press or through the use of a large multi-color press, will normally be sufficient to produce all of the final colors desired. The four standard printing colors are process yellow, a slightly greenish yellow; process magenta, a red-purple; process cyan, a slightly greenish-blue; and black. When four colors other than those specified above are desired, the printer can mix inks to conform to the desired colors. Using fewer than four colors is also an option. For example, if only black were omitted, there is still a tremendous range of possibilities with yellow, magenta, and cyan. If black and cyan were both eliminated, it would still be possible to achieve a color scheme consisting of yellow, yellow-orange, orange, red-orange, and magenta. If black and magenta were both dispensed with, it would still be possible to achieve a color scheme consisting of yellow, yellow-green, green, and blue-green.

The color-mixing capabilities of modern computer hardware also offer a wide choice of shades and patterns. Although the mechanics differ from printing processes, the same principles of color selection are followed.

The Number of Classes Required

The more classes that are required, the greater the benefits provided by color, but likewise, establishing the color scheme will become increasingly complex. For five or fewer classes a distinctive hue can be used for each class with little difficulty.

If gray is acceptable, six classes could be achieved. For more than five classes, it will be necessary either to use tertiary hues (which are harder to differentiate, particularly if the symbolism is narrow or small) or to employ the same hue at two or more levels of darkness. With tertiary hues, up to nine classes can be achieved with a separate hue for each—or ten, if gray is acceptable. For more than ten classes, it is essential that the same hue be employed at two or more levels of darkness. In general, the smaller the number of classes, the larger the darkness steps that can be made (assuming that progressive darkness is to be used—which is desirable).

The Type of Symbolism to Employ

The particular symbolism to be employed will prove to be critical. Usually one should select the type of symbolism and the color scheme at the same time. Color may assist materially with certain symbolisms, such as when some of the data zones are especially small and FIELD symbolism is to be employed; when BAND symbolism must be extremely narrow; or when SPOT symbolism is particularly small. In all such cases, high intensity colors will usually be desirable. With FIELD symbolism containing large data zones, or when SPOT or BAND symbolisms are used over large areas, more subtle shades of lower intensity will usually be preferable. Symbolism of 3-dimensional type or symbolism based on countability will typically require darker tones to be successful.

Color by Number of Classes

After considering the rather special problem of two classes, we will describe the use of primary and secondary hues only, with data assumed to be grouped in a progressively larger number of classes. Next we will deal with the problems of adding tertiary hues. The more classes, the more confusion, hence the greater the hue spread needed. In all cases we have given general guidelines rather than actual color specifications. Whether you use colored pencils, four-color printing, or a color CRT screen, the goals are the same, even though the precise tones available may vary.

Two Classes   Color is never needed to differentiate two classes—hence it is used only for aesthetic reasons. When so used, it is best not to employ radically different hues or tones. Almost any two tones may be used, provided that they are somewhat related yet easily distinguishable. One of the tones must, however, clearly suggest a larger quantity than the other. This goal can usually be achieved most effectively by making one color substantially darker than the other, or substantially more intense. With only two hues, it is usually necessary that they not be separated by more than one step or thirty degrees on the color circle, to achieve a sense of relatedness.

When color areas are relatively large, there is a risk that high intensity colors will appear crude or harsh. Consequently, it may be desirable to use two tones of equally low intensity or, preferably, a low intensity tone and a darker tone which is somewhat more intense.

On the warm side, the best choices might be orange and red-orange, or red-orange and red. On the cool side, comparable choices might be yellow-green and green, green and blue-green, blue-green and blue, or blue and blue-purple. The more dominant warm combinations are preferable if the areas of color will be relatively small. If the areas of color will be relatively large, the cool combinations might prove less harsh and therefore preferable from a purely aesthetic viewpoint.

Three Classes For three classes, it is best to choose hues separated by sixty degrees or more on the color circle. On the warm side the most obvious choice would be yellow, orange, and red, or possibly yellow-orange, red-orange, and red-purple, provided the last was not less dominant than the red-orange. The first-mentioned solution of yellow, orange, red is probably preferable under most circumstances.

On the cool side, the best choice would be yellow, green, and blue. Less desirable choices are yellow-green, blue-green, and blue-purple—or even green, blue, and purple, if the purple were more dominant than the blue.

We must next decide what darkness levels to employ for each of these hues. Ordinarily we would desire the maximum darkness spread possible. However, with only three classes involved, it would be better to adopt a more restricted darkness spread, with a strong but not overly dark red.

If high intensity had been unnecessary or undesirable, the same three darkness levels could be employed at lower intensity. Remember that the lower the intensity, the less the visual differentiability between adjacent hues. However, when low intensity is employed the contrast in darkness can be increased.

For the preferred sequence on the cool side of the color circle—yellow, green, and blue—the case is similar. If we wished to use lower intensity in general, the same color sequence could be employed with the intensity reduced as desired, though not so much as to make the tones difficult to distinguish or hard to name. A lower intensity would probably be successful with blue and green, since they would remain readily identifiable.

A different solution is to use the three primary colors—with the intensity of blue reduced to place it at the halfway point between yellow and red. While the three colors are unrelated in hue, they are obviously differentiable and easy to name. In this case red, the most dominant of all colors, is the terminal color.

If gray is to be used as a quantitative analogue it should, for the reasons previously explained, be employed as the lightest tone, at a darkness percentage probably not less than 10 percent. The remaining two tones, both of positive hue, might best be yellow-orange and red-orange on the warm side, or yellow-green and blue-green on the cool side, with the darkness level of the intermediate tone in each case falling about midway between that of the gray and the terminal tone. Alternative sequences might consist of gray, orange, red; or gray, green, blue; but the tertiary hues of yellow-orange or yellow-green seem to provide a greater sense of continuity with gray.

Four Classes As before, we will start by assuming that high intensity is desired—since this is likely to be the more restrictive case.

Unless gray is to be employed, no solution is possible on the warm side of the color circle without either using tertiaries, violating the recommendations given as to dominance, or using the same hue for more than one class.

On the cool side, a sequence of yellow, green, blue, purple could be employed. Yellow might be set at approximately 20 percent darkness and the purple terminal position at 90 percent darkness, the maximum desirable. Blue would be slightly less than 100 percent intensity, green somewhat less than 75 percent, and yellow 85 percent.

With four classes, another possibility merits consideration—that of employing the same hue at more than one darkness level. This could have been done in the case of three classes, but hardly seemed justified. It may not prove significant until six or more classes are involved.

We start by selecting two hues as discussed for two classes, and then duplicate each at successive darkness levels to yield four tones. This can be done photographically by screening a color at varying degrees of darkness or by manipulating dot combinations on a color CRT unit. An increase in darkness of a light hue may also be achieved manually by superimposing a somewhat coarse pattern of black dots, lines, or other shapes. Or we might select two relatively dark hues and duplicate each at a lighter level by superimposing a pattern of white dots, lines, or other shapes. With either of these procedures, it will be desirable to employ hues of relatively high intensity. The object in using a coarse pattern is to increase to the utmost differentiability between tones of the same hue. Without such superimposed patterns, the difference in darkness would be relatively small. (When using overlaid patterns, note that intensity drops off with greater than normal rapidity, owing to the reduction of exposed hue.)

If superimposed black patterns are to be employed for the darker tone of each pair, start with two relatively light hues, such as high-intensity yellow and orange, or yellow and green. With these hues the straight yellow can be raised to 20 percent darkness and the straight orange to 60 percent darkness. (Both hues when "straight" would be of 100 percent intensity.) A similar solution could be used for the cool side.

If superimposed white patterns are to be employed for the lighter tone of each pair, we must start with two relatively dark hues. For this purpose the pairs originally suggested (orange and red, green and blue) should work very successfully, though greater darkness steps will be desirable. For straight orange, a darkness of 90 percent with a relative intensity of about 90 percent is suggested. This yields darkness steps of 20 percent, which should work well with hue or overlay change. On the cool side, green at about 85 percent relative intensity should prove adequate.

In general, black dots or other overlays will be most successful with tones of about 60 percent darkness and lighter, and white dots will be most successful with tones of about 50 percent darkness and darker. Light yellow or any other light hue (necessarily of low relative intensity) should not normally be used with white overlay, and dark purple or any other dark hue (of high or low intensity) should not normally be used with black overlay.

Five Classes   Since five classes will probably be used most frequently, the planning of color for five classes warrants special consideration.

With the constraints of a separation of at least sixty degrees between hues and the use of each hue no more than once, no solution can be found which lies entirely on either the warm or cool side of the color circle. With the addition of gray, the following possibility is found on the cool side: gray, yellow, green, blue, purple. This is, however, less than ideal—as purple is the terminal hue.

The best possibility is to make use of both the cool and warm sides of the color circle by employing yellow, green, blue, purple, and red. This solution does depend on the use of purple, but purple is not the terminal hue and since it relates to the surrounding blue and red, the problems are minimized. Care must be taken, however, not to use a purple that is darker than the red or less intense than the blue.

Employing the same hue at two or more darkness levels, subject to the options previously considered, yields several acceptable solutions: yellow, orange, red; or yellow, green, blue—with the last two in each series employed at two darkness levels. Gray could be substituted for yellow in these two schemes, but if the four-color process were used, there would be no savings in doing so. If gray were to be employed at two levels of darkness, the single positive hue could be used in either the intermediate or terminal position. In a color scheme composed of gray, darker gray, orange, darker orange, and red, the fifth and darkest color would be particularly emphasized. In contrast, if there were only one class of orange and two classes of red, the last two would both be emphasized due to the dominant character of red. There is less distinction between the two alternatives on the cool side, as there is less difference in dominance between green and blue than between orange and red.

A final alternative, assuming the use of gray and making use of both the cool and warm sides of the color circle, is the following: gray, yellow-green, blue-green, blue-purple, red-purple. One advantage of this sequence is the elimination of yellow, a weak color against a white background. This scheme might be undesirable, however, because of the difficulty in procuring accurate tertiary colors and the fact that these colors are less easily named.

Six Classes   Assuming sixty degrees separation on the color circle, there is no acceptable solution for six classes without using gray. This is because six hues are the maximum that can fit within the color circle at sixty-degree spacing, and every possible darkness sequence would violate the dominance recommendations.

If gray is used, the problem is essentially similar to that of five classes, except for the addition of gray to produce the sequence gray, yellow, green, blue, purple, red. However, as it is usually wise to employ gray at less than 10 percent darkness, it may be desirable to make minor adjustments of darkness in relation to the other tones.

Assuming each hue is employed in two levels of darkness, we may use the hue sequences suited to three classes: yellow, orange, red; and yellow, green, blue. Gray could be substituted for yellow if desired, though in terms of the four-color process, yellow would still be required to produce the orange or green. In the last-mentioned sequences, instead of using yellow at two darkness levels (which would present difficulties with overlaid white dots), gray and one darkness level of yellow could be used for the first two classes.

Seven Classes   When hues are separated by only thirty degrees, tertiary hues must be employed, and a high degree of accuracy is necessary. When the colors are in the form of ready-mixed ink or self-sticking plastic sheets, they may be tested before use and freely employed if satisfactory. With the four-color process, however, it is difficult to assure satisfactory results under widely varying production conditions, and in consequence special precautions should be taken.

Other Options   One approach that is helpful when a large number of value levels is required is to preserve the advantages of progressive darkness without requiring strict overall progression. Based on the use of several hues of progressive darkness, this procedure achieves several darkness levels within each hue by means of shaped-tone symbolism. The hues should be primary, or primary and secondary, and of full intensity to optimize differentiability. As many as five or six darkness levels can be successfully employed with each hue. Thus, with only three hues, it would be possible to have fifteen or eighteen classes; or with five hues, twenty-five or thirty classes.

As was already noted, the tones for these classes cannot be of continuously increasing darkness. When examining such a map, the user would first become aware of the main categories based upon hues of progressive darkness before proceeding to the individual value levels based upon the increasing darknesses within each hue. To judge the latter with accuracy, the user would be helped by the changes in shape provided by the use of shaped-tone symbolism within each hue.

In choosing colors to serve as quantitative analogues, the best general guide is to combine constant or, preferably, increasing intensity with increasing darkness as larger values are represented. When color is used for isolated dots or lines (as in SPOT or BAND symbolism), it is important that the dots be large enough and the lines wide enough so that the hue can be read effectively. Dots and lines, unless of considerable size, should not be placed so close together that the combination of hues involved produces colors other than those intended. Further, colors employed to differentiate subjects of equal importance must be of approximately the same darkness and intensity.

The preceding discussion is meant to highlight those issues and principles that are valid in the use of color for quantitative mapping. It must be recognized, however, that there is great potential for variety and that continued research and experimentation are highly desirable and will probably be rewarding.

Appendices

APPENDIX 1

Using Reciprocal Curve Classing

To use reciprocal curve classing, follow the procedure outlined below.

STEP 1

On a sheet of transparent paper, trace the outline of the chart in Figure 10–15. Then, placing the same sheet over a plain white surface, plot the value curve for the particular value set to be displayed. (Follow the procedure illustrated in the large upper chart of Figure 10–3, using the dot at the lower left corner to represent the minimum value and the dot at the upper right corner for the maximum value. All other values are located between these, according to increasing magnitude from left to right.)

If the value curve thus drawn proves to be reasonably straight in general trend, equal class spans will be preferable in most cases. If the value curve is not reasonably straight, replace the transparent paper over Figure 10–15 and, ignoring the dashed portions of the curves, select that classing curve that best represents the trend of the value curve. If the trend of the value curve conforms only roughly to any classing curve shown, select that classing curve about which the value curve appears to be most evenly balanced. If the value curve falls between two classing curves shown, roughly interpolate between them by eye. With some idea of a good classing curve to employ, turn to Table A1–1, at the end of this Appendix.

STEP 2a

If the curve selected is one of those illustrated, turn to the applicable cluster of figures given in the table. For example, if curve 45 is selected, find the cluster of figures identified by the words "RECIP- 45.0." If only two classes are to be employed, the first figure given at the left (23.68) in the column headed "2" at the top of the sheet indicates (in terms of percentage between the value extremes) the applicable breakpoint—i.e., the lowest value for the second class. (The highest value of the first class would be the same figure with a "less than" sign before it.) From this figure, the *tentative* class spans (in terms of percentage) can be immediately established. The final class spans are then determined from these percentage figures. If the value range were 300–400, the difference (100) would represent 100 percent, and the class breakpoint would be established by taking 23.68 percent of 100 and adding the result to the lower figure (300) for a final classing of 300–<323.68 and 323.68–400. If the value range happened to be 41–87, the difference (46) would represent 100 percent, and the class breakpoint would be established by taking 23.68 percent of 46 and adding the result to the lower figure (41) for a final classing of 41–<51.89 and 51.89–87. If less ragged figures were desired, the *final* figures might be rounded to nearest integer values. If more than two classes are to be employed, use the corresponding figures that appear in the columns toward the right; for five classes, use the figures in the fourth column (headed "5" at the top of the sheet), and so on. In each case, the figures given show the lowest value in each class after the first, in terms of percentage between the value extremes.

With a high value curve (those curves labeled "inverted" on the chart), follow the same procedure except that the figures given in the table must be subtracted from 100. Thus, if curve I–45 were selected, the percentage values for two classes would be 0–<76.32 and 76.32–100.

STEP 2b

The following procedure should be used when the curve selected is not in the classing curve chart. In the extreme left-hand column of Figure A1–1, under "curves plotted," find the numbers of the two curves between which the selected curve falls. Then, pursuant to the earlier suggested interpolation, select from the adjacent column that curve which appears most appropriate. Turn to the corresponding cluster of figures in Table A1–1 and proceed as previously described.

The method used in establishing the number of location per class is as follows:

STEP 3

As in Figure 10–16A, at each class breakpoint, thin vertical lines are drawn so as to fall midway horizontally between the highest dot in the next lower class and the lowest dot in the next higher class. Short vertical lines are then drawn below and slightly outside the lower corners of the chart, at a distance equal to half the regular horizontal spacing between dots. The number of dots between the vertical lines is the number of locations per class. Distances between the vertical lines correspond to the heights of the upper bars of Figures 10–16B and 10–16C.

Such a chart can be made for any given set of values regardless of the shape of the value curve, and using any given classing system.

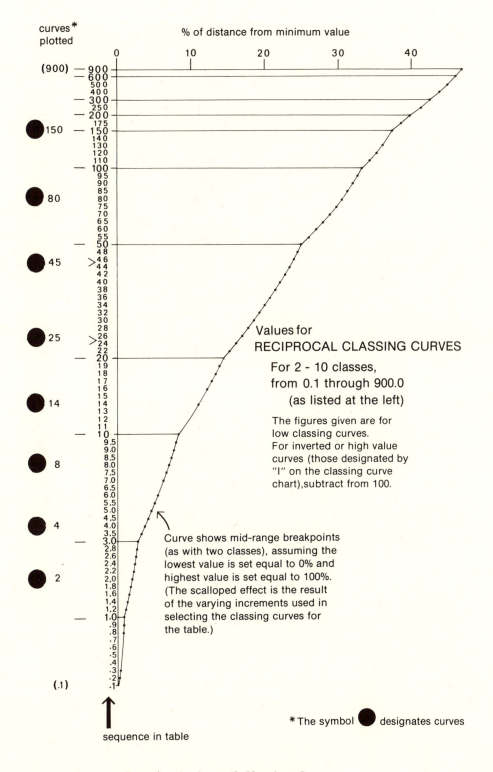

Figure A1–1 Values for Reciprocal Classing Curves

Table A1–1: Reciprocal Curves—Lower Limits of Classes

				Number of classes–				
2	3	4	5	6	7	8	9	10
0.10	0.05	0.03	0.02	0.02	0.02	0.01	0.01	0.01
	0.20	0.10	0.07	0.05	0.04	0.03	0.03	0.02
		0.30	0.15	0.10	0.07	0.06	0.05	0.04
			0.40	0.20	0.13	0.10	0.08	0.07
Recip–	0.1			0.50	0.25	0.17	0.12	0.10
					0.60	0.30	0.20	0.15
						0.69	0.35	0.23
							0.79	0.40
								0.89
0.20	0.10	0.07	0.05	0.04	0.03	0.03	0.02	0.02
	0.40	0.20	0.13	0.10	0.08	0.07	0.06	0.05
		0.60	0.30	0.20	0.15	0.12	0.10	0.09
			0.79	0.40	0.27	0.20	0.16	0.13
Recip–	0.2			0.99	0.50	0.33	0.25	0.20
					1.18	0.60	0.40	0.30
						1.38	0.69	0.46
							1.57	0.79
								1.76
0.30	0.15	0.10	0.07	0.06	0.05	0.04	0.04	0.03
	0.59	0.30	0.20	0.15	0.12	0.10	0.09	0.07
		0.89	0.45	0.30	0.22	0.18	0.15	0.13
			1.18	0.59	0.40	0.30	0.24	0.20
Recip–	0.3			1.47	0.74	0.50	0.37	0.30
					1.76	0.89	0.59	0.45
						2.05	1.04	0.69
							2.34	1.18
								2.62
0.40	0.20	0.13	0.10	0.08	0.07	0.06	0.05	0.04
	0.79	0.40	0.26	0.20	0.16	0.13	0.11	0.10
		1.18	0.59	0.40	0.30	0.24	0.20	0.17
			1.57	0.79	0.53	0.40	0.32	0.26
Recip–	0.4			1.95	0.99	0.66	0.50	0.40
					2.33	1.18	0.79	0.59
						2.71	1.38	0.92
							3.09	1.57
								3.46
0.50	0.25	0.17	0.12	0.10	0.08	0.07	0.06	0.06
	0.99	0.50	0.33	0.25	0.20	0.17	0.14	0.12
		1.47	0.74	0.50	0.37	0.30	0.25	0.21
			1.95	0.99	0.66	0.50	0.40	0.33
Recip–	0.5			2.43	1.23	0.82	0.62	0.50
					2.90	1.47	0.99	0.74
						3.37	1.71	1.15
							3.83	1.95
								4.29

Table A1–1: *(continued)*

			Number of classes–					
2	3	4	5	6	7	8	9	10
0.59	0.30	0.20	0.15	0.12	0.10	0.09	0.07	0.07
	1.18	0.59	0.40	0.30	0.24	0.20	0.17	0.15
		1.76	0.89	0.59	0.45	0.36	0.30	0.25
			2.33	1.18	0.79	0.59	0.47	0.40
Recip–	0.6			2.90	1.47	0.98	0.74	0.59
					3.45	1.76	1.18	0.89
						4.01	2.05	1.37
							4.55	2.33
								5.09
0.69	0.35	0.23	0.17	0.14	0.12	0.10	0.09	0.08
	1.37	0.69	0.46	0.35	0.28	0.23	0.20	0.17
		2.04	1.03	0.69	0.52	0.42	0.35	0.30
			2.71	1.37	0.92	0.69	0.55	0.46
Recip–	0.7			3.36	1.71	1.15	0.86	0.69
					4.00	2.04	1.37	1.03
						4.64	2.38	1.60
							5.27	2.71
								5.89
0.79	0.40	0.26	0.20	0.16	0.13	0.11	0.10	0.09
	1.56	0.79	0.53	0.40	0.32	0.26	0.23	0.20
		2.33	1.18	0.79	0.59	0.47	0.40	0.34
			3.08	1.56	1.05	0.79	0.63	0.53
Recip–	0.8			3.82	1.95	1.31	0.98	0.79
					4.54	2.33	1.56	1.18
						5.26	2.70	1.82
							5.97	3.08
								6.67
0.88	0.44	0.30	0.22	0.18	0.15	0.13	0.11	0.10
	1.75	0.88	0.59	0.44	0.36	0.30	0.25	0.22
		2.61	1.32	0.88	0.66	0.53	0.44	0.38
			3.44	1.75	1.18	0.88	0.71	0.59
Recip–	0.9			4.27	2.18	1.46	1.10	0.88
					5.08	2.61	1.75	1.32
						5.88	3.03	2.04
							6.66	3.44
								7.43
0.98	0.49	0.33	0.25	0.20	0.16	0.14	0.12	0.11
	1.94	0.98	0.66	0.49	0.39	0.33	0.28	0.25
		2.88	1.46	0.98	0.74	0.59	0.49	0.42
			3.81	1.94	1.30	0.98	0.79	0.66
Recip–	1.0			4.72	2.42	1.62	1.22	0.98
					5.61	2.88	1.94	1.46
						6.48	3.35	2.26
							7.34	3.81
								8.18

Table A1–1: *(continued)*

	2	3	4	5	6	7	8	9	10
				Number of classes–					
	1.17	0.59	0.39	0.30	0.24	0.20	0.17	0.15	0.13
		2.32	1.17	0.78	0.59	0.47	0.39	0.34	0.30
			3.44	1.75	1.17	0.88	0.71	0.59	0.51
				4.53	2.32	1.56	1.17	0.94	0.78
	Recip–	1.2			5.60	2.88	1.94	1.46	1.17
						6.64	3.44	2.32	1.75
							7.66	3.99	2.69
								8.67	4.53
									9.64
	1.36	0.69	0.46	0.34	0.28	0.23	0.20	0.17	0.15
		2.69	1.36	0.91	0.69	0.55	0.46	0.39	0.34
			3.98	2.03	1.36	1.02	0.82	0.69	0.59
				5.23	2.69	1.81	1.36	1.09	0.91
	Recip–	1.4			6.46	3.34	2.25	1.70	1.36
						7.65	3.98	2.69	2.03
							8.81	4.61	3.12
								9.95	5.23
									11.05
	1.55	0.78	0.52	0.39	0.31	0.26	0.22	0.20	0.17
		3.05	1.55	1.04	0.78	0.63	0.52	0.45	0.39
			4.51	2.31	1.55	1.17	0.94	0.78	0.67
				5.93	3.05	2.06	1.55	1.24	1.04
	Recip–	1.6			7.30	3.79	2.56	1.93	1.55
						8.63	4.51	3.05	2.31
							9.93	5.22	3.54
								11.19	5.93
									12.41
	1.74	0.88	0.59	0.44	0.35	0.29	0.25	0.22	0.20
		3.42	1.74	1.17	0.88	0.70	0.59	0.50	0.44
			5.04	2.58	1.74	1.31	1.05	0.88	0.75
				6.61	3.42	2.30	1.74	1.39	1.17
	Recip–	1.8			8.12	4.23	2.86	2.16	1.74
						9.59	5.04	3.42	2.58
							11.01	5.83	3.96
								12.39	6.61
									13.73
	1.92	0.97	0.65	0.49	0.39	0.33	0.28	0.24	0.22
		3.77	1.92	1.29	0.97	0.78	0.65	0.56	0.49
			5.56	2.86	1.92	1.45	1.16	0.97	0.83
				7.27	3.77	2.55	1.92	1.54	1.29
	Recip–	2.0			8.93	4.67	3.16	2.39	1.92
						10.52	5.56	3.77	2.86
							12.07	6.42	4.38
								13.56	7.27
									15.00

Table A1–1: *(continued)*

				Number of classes–				
2	3	4	5	6	7	8	9	10
2.11	1.06	0.71	0.54	0.43	0.36	0.31	0.27	0.24
	4.13	2.11	1.41	1.06	0.85	0.71	0.61	0.54
		6.07	3.13	2.11	1.59	1.28	1.06	0.91
			7.93	4.13	2.79	2.11	1.69	1.41
Recip–	2.2			9.72	5.11	3.46	2.62	2.11
					11.43	6.07	4.13	3.13
						13.10	7.01	4.78
							14.69	7.93
								16.23
2.29	1.16	0.78	0.58	0.47	0.39	0.33	0.29	0.26
	4.48	2.29	1.54	1.16	0.93	0.78	0.67	0.58
		6.57	3.40	2.29	1.73	1.39	1.16	0.99
			8.57	4.48	3.03	2.29	1.84	1.54
Recip–	2.4			10.49	5.54	3.76	2.85	2.29
					12.32	6.57	4.48	3.40
						14.09	7.58	5.19
							15.79	8.57
								17.42
2.47	1.25	0.84	0.63	0.50	0.42	0.36	0.32	0.28
	4.82	2.47	1.66	1.25	1.00	0.84	0.72	0.63
		7.07	3.66	2.47	1.86	1.50	1.25	1.07
			9.20	4.82	3.27	2.47	1.99	1.66
Recip–	2.6			11.24	5.96	4.05	3.07	2.47
					13.19	7.07	4.82	3.66
						15.07	8.15	5.58
							16.86	9.20
								18.57
2.65	1.34	0.90	0.68	0.54	0.45	0.39	0.34	0.30
	5.17	2.65	1.78	1.34	1.08	0.90	0.77	0.68
		7.55	3.93	2.65	2.00	1.61	1.34	1.15
			9.82	5.17	3.50	2.65	2.13	1.78
Recip–	2.8			11.98	6.38	4.34	3.29	2.65
					14.04	7.55	5.17	3.93
						16.01	8.70	5.98
							17.89	9.82
								19.69
2.83	1.44	0.96	0.72	0.58	0.48	0.41	0.36	0.32
	5.51	2.83	1.90	1.44	1.15	0.96	0.83	0.72
		8.04	4.19	2.83	2.14	1.72	1.44	1.23
			10.43	5.51	3.74	2.83	2.28	1.90
Recip–	3.0			12.71	6.79	4.63	3.51	2.83
					14.87	8.04	5.51	4.19
						16.94	9.25	6.36
							18.90	10.43
								20.77

Table A1–1: *(continued)*

				Number of classes–				
2	3	4	5	6	7	8	9	10
3.27	1.66	1.11	0.84	0.67	0.56	0.48	0.42	0.37
	6.34	3.27	2.20	1.66	1.33	1.11	0.96	0.84
		9.21	4.83	3.27	2.47	1.99	1.66	1.43
			11.91	6.34	4.31	3.27	2.63	2.20
Recip–	3.5			14.46	7.80	5.34	4.06	3.27
					16.86	9.21	6.34	4.83
						19.14	10.58	7.31
							21.29	11.91
								23.33
3.70	1.89	1.27	0.95	0.76	0.64	0.55	0.48	0.43
	7.14	3.70	2.50	1.89	1.52	1.27	1.09	0.95
		10.34	5.45	3.70	2.80	2.26	1.89	1.62
			13.33	7.14	4.88	3.70	2.98	2.50
Recip–	4.0			16.13	8.77	6.02	4.59	3.70
					18.74	10.34	7.14	5.45
						21.21	11.87	8.24
							23.53	13.33
								25.71
4.13	2.11	1.42	1.07	0.85	0.71	0.61	0.54	0.48
	7.93	4.13	2.79	2.11	1.69	1.42	1.22	1.07
		11.44	6.07	4.13	3.13	2.52	2.11	1.81
			14.69	7.93	5.43	4.13	3.33	2.79
Recip–	4.5			17.71	9.72	6.70	5.11	4.13
					20.53	11.44	7.93	6.07
						23.16	13.10	9.13
							25.62	14.69
								27.93
4.55	2.33	1.56	1.18	0.94	0.79	0.68	0.59	0.53
	8.70	4.55	3.08	2.33	1.87	1.56	1.34	1.18
		12.50	6.67	4.55	3.45	2.78	2.33	2.00
			16.00	8.70	5.97	4.55	3.67	3.08
Recip–	5.0			19.23	10.64	7.35	5.62	4.55
					22.22	12.50	8.70	6.67
						25.00	14.29	10.00
							27.59	16.00
								30.00
4.95	2.54	1.71	1.29	1.03	0.86	0.74	0.65	0.58
	9.44	4.95	3.36	2.54	2.04	1.71	1.47	1.29
		13.52	7.25	4.95	3.76	3.03	2.54	2.19
			17.25	9.44	6.50	4.95	4.00	3.36
Recip–	5.5			20.67	11.53	7.99	6.12	4.95
					23.82	13.52	9.44	7.25
						26.74	15.43	10.85
							29.43	17.25
								31.94

Table A1–1: *(continued)*

	Number of classes–								
2	3	4	5	6	7	8	9	10	
5.36	2.75	1.85	1.40	1.12	0.93	0.80	0.70	0.62	
	10.17	5.36	3.64	2.75	2.21	1.85	1.59	1.40	
		14.52	7.83	5.36	4.07	3.28	2.75	2.37	
			18.46	10.17	7.02	5.36	4.33	3.64	
Recip–	6.0			22.05	12.40	8.62	6.61	5.36	
					25.35	14.52	10.17	7.83	
						28.38	16.54	11.67	
							31.17	18.46	
								33.75	
5.75	2.96	1.99	1.50	1.21	1.01	0.86	0.76	0.67	
	10.88	5.75	3.91	2.96	2.38	1.99	1.71	1.50	
		15.48	8.39	5.75	4.38	3.53	2.96	2.55	
			19.62	10.88	7.52	5.75	4.65	3.91	
Recip–	6.5			23.38	13.24	9.23	7.09	5.75	
					26.80	15.48	10.88	8.39	
						29.93	17.60	12.47	
							32.81	19.62	
								35.45	
6.14	3.17	2.13	1.61	1.29	1.08	0.93	0.81	0.72	
	11.57	6.14	4.18	3.17	2.55	2.13	1.83	1.61	
		16.41	8.94	6.14	4.68	3.78	3.17	2.73	
			20.74	11.57	8.02	6.14	4.97	4.18	
Recip–	7.0			24.64	14.06	9.83	7.56	6.14	
					28.18	16.41	11.57	8.94	
						31.41	18.63	13.24	
							34.36	20.74	
								37.06	
6.52	3.37	2.27	1.71	1.38	1.15	0.99	0.86	0.77	
	12.25	6.52	4.44	3.37	2.71	2.27	1.95	1.71	
		17.31	9.47	6.52	4.97	4.02	3.37	2.90	
			21.82	12.25	8.51	6.52	5.29	4.44	
Recip–	7.5			25.86	14.85	10.42	8.02	6.52	
					29.50	17.31	12.25	9.47	
						32.81	19.63	14.00	
							35.82	21.82	
								38.57	
6.90	3.57	2.41	1.82	1.46	1.22	1.05	0.92	0.82	
	12.90	6.90	4.71	3.57	2.88	2.41	2.07	1.82	
		18.18	10.00	6.90	5.26	4.26	3.57	3.08	
			22.86	12.90	8.99	6.90	5.59	4.71	
Recip–	8.0			27.02	15.63	10.99	8.48	6.90	
					30.76	18.18	12.90	10.00	
						34.15	20.59	14.74	
							37.21	22.86	
								40.00	

Table A1–1: *(continued)*

| | | | | *Number of classes–* | | | | |
2	3	4	5	6	7	8	9	10
7.26	3.77	2.54	1.92	1.54	1.29	1.11	0.97	0.86
	13.55	7.26	4.96	3.77	3.04	2.54	2.19	1.92
		19.03	10.52	7.26	5.55	4.49	3.77	3.25
			23.86	13.55	9.46	7.26	5.90	4.96
Recip–	8.5			28.14	16.38	11.55	8.92	7.26
					31.97	19.03	13.55	10.52
						35.42	21.52	15.45
							38.53	23.86
								41.35
7.63	3.96	2.68	2.02	1.62	1.36	1.17	1.02	0.91
	14.18	7.63	5.22	3.96	3.20	2.68	2.30	2.02
		19.85	11.02	7.63	5.83	4.72	3.96	3.42
			24.83	14.18	9.92	7.63	6.20	5.22
Recip–	9.0			29.22	17.11	12.10	9.36	7.63
					33.12	19.85	14.18	11.02
						36.63	22.42	16.15
							39.78	24.83
								42.63
7.98	4.16	2.81	2.12	1.71	1.42	1.22	1.07	0.95
	14.79	7.98	5.47	4.16	3.35	2.81	2.42	2.12
		20.65	11.52	7.98	6.11	4.95	4.16	3.58
			25.76	14.79	10.37	7.98	6.49	5.47
Recip–	9.5			30.25	17.82	12.63	9.79	7.98
					34.23	20.65	14.79	11.52
						37.78	23.29	16.84
							40.97	25.76
								43.85
8.33	4.35	2.94	2.22	1.79	1.49	1.28	1.12	1.00
	15.39	8.33	5.71	4.35	3.51	2.94	2.53	2.22
		21.43	12.00	8.33	6.38	5.17	4.35	3.75
			26.67	15.39	10.81	8.33	6.78	5.71
Recip–	10.0			31.24	18.52	13.16	10.21	8.33
					35.29	21.43	15.39	12.00
						38.89	24.14	17.50
							42.11	26.67
								45.00
9.02	4.72	3.20	2.42	1.94	1.62	1.40	1.22	1.09
	16.54	9.02	6.20	4.72	3.81	3.20	2.75	2.42
		22.92	12.94	9.02	6.92	5.61	4.72	4.07
			28.39	16.54	11.67	9.02	7.34	6.20
Recip–	11.0			33.13	19.86	14.18	11.02	9.02
					37.28	22.92	16.54	12.94
						40.96	25.75	18.78
							44.22	28.39
								47.14

Table A1–1: *(continued)*

	Number of classes–								
2	3	4	5	6	7	8	9	10
9.68	5.08	3.45	2.61	2.10	1.75	1.51	1.32	1.18
	17.65	9.68	6.67	5.08	4.11	3.45	2.97	2.61
		24.32	13.85	9.68	7.44	6.04	5.08	4.39
			30.00	17.65	12.50	9.68	7.89	6.67
Recip–	12.0			34.88	21.13	15.15	11.81	9.68
					39.12	24.32	17.65	13.85
						42.86	27.28	20.00
							46.16	30.00
								49.09
10.32	5.44	3.69	2.80	2.25	1.88	1.62	1.42	1.26
	18.71	10.32	7.12	5.44	4.40	3.69	3.18	2.80
		25.66	14.72	10.32	7.94	6.46	5.44	4.70
			31.52	18.71	13.30	10.32	8.43	7.12
Recip–	13.0			36.51	22.34	16.09	12.57	10.32
					40.83	25.66	18.71	14.72
						44.61	28.71	21.16
							47.93	31.52
								50.87
10.94	5.78	3.93	2.98	2.40	2.00	1.72	1.51	1.35
	19.72	10.94	7.57	5.78	4.68	3.93	3.39	2.98
		26.92	15.56	10.94	8.43	6.86	5.78	5.00
			32.94	19.72	14.07	10.94	8.94	7.57
Recip–	14.0			38.04	23.49	16.99	13.31	10.94
					42.42	26.92	19.72	15.56
						46.23	30.06	22.27
							49.56	32.94
								52.50
11.54	6.12	4.17	3.16	2.54	2.13	1.83	1.60	1.43
	20.69	11.54	8.00	6.12	4.96	4.17	3.59	3.16
		28.12	16.36	11.54	8.91	7.26	6.12	5.29
			34.29	20.69	14.81	11.54	9.45	8.00
Recip–	15.0			39.47	24.59	17.86	14.02	11.54
					43.89	28.12	20.69	16.36
						47.73	31.35	23.33
							51.07	34.29
								54.00
12.12	6.45	4.40	3.33	2.69	2.25	1.93	1.69	1.51
	21.62	12.12	8.42	6.45	5.23	4.40	3.79	3.33
		29.27	17.14	12.12	9.37	7.64	6.45	5.58
			35.56	21.62	15.53	12.12	9.94	8.42
Recip–	16.0			40.81	25.64	18.69	14.71	12.12
					45.27	29.27	21.62	17.14
						49.12	32.56	24.35
							52.46	35.56
								55.38

Table A1–1: *(continued)*

			Number of classes–					
2	3	4	5	6	7	8	9	10
12.69	6.77	4.62	3.51	2.82	2.36	2.03	1.78	1.59
	22.52	12.69	8.83	6.77	5.49	4.62	3.99	3.51
		30.36	17.89	12.69	9.82	8.02	6.77	5.86
			36.76	22.52	16.23	12.69	10.41	8.83
Recip–	17.0			42.07	26.65	19.50	15.37	12.69
					46.57	30.36	22.52	17.89
						50.42	33.71	25.32
							53.76	36.76
								56.67
13.24	7.09	4.84	3.67	2.96	2.48	2.13	1.87	1.67
	23.38	13.24	9.23	7.09	5.75	4.84	4.18	3.67
		31.40	18.62	13.24	10.26	8.39	7.09	6.14
			37.89	23.38	16.90	13.24	10.87	9.23
Recip–	18.0			43.26	27.61	20.27	16.02	13.24
					47.78	31.40	23.38	18.62
						51.64	34.81	26.25
							54.96	37.89
								57.86
13.77	7.39	5.05	3.84	3.10	2.59	2.23	1.96	1.74
	24.21	13.77	9.62	7.39	6.00	5.05	4.36	3.84
		32.39	19.32	13.77	10.69	8.74	7.39	6.40
			38.97	24.21	17.55	13.77	11.32	9.62
Recip–	19.0			44.39	28.53	21.02	16.64	13.77
					48.92	32.39	24.21	19.32
						52.78	35.85	27.14
							56.09	38.97
								58.97
14.29	7.69	5.26	4.00	3.23	2.70	2.33	2.04	1.82
	25.00	14.29	10.00	7.69	6.25	5.26	4.54	4.00
		33.33	20.00	14.29	11.11	9.09	7.69	6.67
			40.00	25.00	18.18	14.29	11.76	10.00
Recip–	20.0			45.45	29.41	21.74	17.24	14.29
					49.99	33.33	25.00	20.00
						53.85	36.85	28.00
							57.15	40.00
								60.00
15.28	8.27	5.67	4.31	3.48	2.92	2.51	2.20	1.96
	26.51	15.28	10.73	8.27	6.73	5.67	4.90	4.31
		35.11	21.29	15.28	11.91	9.76	8.27	7.17
			41.90	26.51	19.38	15.28	12.61	10.73
Recip–	22.0			47.41	31.07	23.11	18.40	15.28
					51.96	35.11	26.51	21.29
						55.80	38.70	29.62
							59.06	41.90
								61.88

Table A1–1: *(continued)*

	Number of classes–								
2	3	4	5	6	7	8	9	10
16.22	8.82	6.06	4.62	3.73	3.12	2.69	2.36	2.11
	27.91	16.22	11.43	8.82	7.19	6.06	5.24	4.62
		36.73	22.50	16.22	12.67	10.40	8.82	7.66
			43.64	27.91	20.51	16.22	13.41	11.43
Recip–	24.0			49.17	32.61	24.39	19.48	16.22
					53.72	36.73	27.91	22.50
						57.53	40.39	31.11
							60.76	43.64
								63.53
16.67	9.09	6.25	4.76	3.85	3.22	2.78	2.44	2.17
	28.57	16.67	11.76	9.09	7.41	6.25	5.40	4.76
		37.50	23.08	16.67	13.04	10.71	9.09	7.89
			44.44	28.57	21.05	16.67	13.79	11.76
Recip–	25.0			49.99	33.33	25.00	20.00	16.67
					54.54	37.50	28.57	23.08
						58.33	41.18	31.82
							61.54	44.44
								64.29
17.11	9.35	6.44	4.91	3.96	3.32	2.86	2.51	2.24
	29.22	17.11	12.09	9.35	7.62	6.44	5.57	4.91
		38.24	23.64	17.11	13.40	11.02	9.35	8.12
			45.22	29.22	21.57	17.11	14.17	12.09
Recip–	26.0			50.78	34.03	25.59	20.51	17.11
					55.31	38.24	29.22	23.64
						59.09	41.94	32.50
							62.28	45.22
								65.00
17.95	9.86	6.80	5.19	4.19	3.52	3.03	2.66	2.37
	30.44	17.95	12.73	9.86	8.05	6.80	5.88	5.19
		39.62	24.71	17.95	14.09	11.60	9.86	8.57
			46.67	30.44	22.58	17.95	14.89	12.73
Recip–	28.0			52.23	35.36	26.72	21.48	17.95
					56.75	39.62	30.44	24.71
						60.49	43.37	33.79
							63.64	46.67
								66.32
18.75	10.34	7.14	5.45	4.41	3.70	3.19	2.80	2.50
	31.58	18.75	13.33	10.34	8.45	7.14	6.18	5.45
		40.91	25.71	18.75	14.75	12.16	10.34	9.00
			48.00	31.58	23.53	18.75	15.58	13.33
Recip–	30.0			53.57	36.59	27.78	22.39	18.75
					58.06	40.91	31.58	25.71
						61.76	44.68	35.00
							64.87	48.00
								67.50

Table A1–1: *(continued)*

				Number of classes–				
2	3	4	5	6	7	8	9	10
19.51	10.81	7.48	5.71	4.63	3.88	3.35	2.94	2.62
	32.66	19.51	13.91	10.81	8.84	7.48	6.48	5.71
		42.11	26.67	19.51	15.38	12.70	10.81	9.41
			49.23	32.66	24.43	19.51	16.24	13.91
Recip–	32.0			54.79	37.74	28.78	23.26	19.51
					59.25	42.11	32.66	26.67
						62.92	45.90	36.13
							65.98	49.23
								68.57
20.24	11.26	7.80	5.96	4.83	4.06	3.50	3.07	2.74
	33.67	20.24	14.47	11.26	9.21	7.80	6.76	5.96
		43.22	27.57	20.24	15.98	13.21	11.26	9.81
			50.37	33.67	25.28	20.24	16.87	14.47
Recip–	34.0			55.92	38.81	29.72	24.08	20.24
					60.35	43.22	33.67	27.57
						63.98	47.04	37.19
							67.00	50.37
								69.55
20.93	11.69	8.11	6.21	5.03	4.22	3.64	3.20	2.86
	34.62	20.93	15.00	11.69	9.57	8.11	7.03	6.21
		44.26	28.42	20.93	16.56	13.71	11.69	10.19
			51.43	34.62	26.08	20.93	17.47	15.00
Recip–	36.0			56.96	39.82	30.61	24.87	20.93
					61.36	44.26	34.62	28.42
						64.95	48.09	38.18
							67.93	51.43
								70.43
21.59	12.10	8.41	6.44	5.22	4.39	3.78	3.33	2.97
	35.52	21.59	15.51	12.10	9.92	8.41	7.29	6.44
		45.24	29.23	21.59	17.11	14.18	12.10	10.56
			52.41	35.52	26.85	21.59	18.05	15.51
Recip–	38.0			57.92	40.77	31.46	25.61	21.59
					62.29	45.24	35.52	29.23
						65.84	49.08	39.12
							68.78	52.41
								71.25
22.22	12.50	8.70	6.67	5.41	4.54	3.92	3.45	3.08
	36.37	22.22	16.00	12.50	10.26	8.70	7.55	6.67
		46.15	30.00	22.22	17.64	14.63	12.50	10.91
			53.33	36.37	27.58	22.22	18.60	16.00
Recip–	40.0			58.82	41.67	32.26	26.32	22.22
					63.15	46.15	36.37	30.00
						66.67	50.00	40.00
							69.57	53.33
								72.00

Table A1–1: *(continued)*

			Number of classes–					
2	3	4	5	6	7	8	9	10
22.83	12.88	8.97	6.89	5.59	4.70	4.05	3.56	3.18
	37.17	22.83	16.47	12.88	10.58	8.97	7.79	6.89
		47.01	30.73	22.83	18.15	15.07	12.88	11.25
			54.19	37.17	28.28	22.83	19.13	16.47
Recip–	42.0			59.65	42.51	33.02	27.00	22.83
					63.95	47.01	37.17	30.73
						67.43	50.87	40.83
							70.30	54.19
								72.69
23.40	13.25	9.25	7.10	5.76	4.84	4.18	3.68	3.28
	37.93	23.40	16.92	13.25	10.89	9.24	8.03	7.10
		47.83	31.43	23.40	18.64	15.49	13.25	11.58
			55.00	37.93	28.94	23.40	19.64	16.92
Recip–	44.0			60.43	43.31	33.74	27.64	23.40
					64.70	47.83	37.93	31.43
						68.14	51.68	41.62
							70.97	55.00
								73.33
23.68	13.43	9.37	7.20	5.85	4.92	4.25	3.73	3.33
	38.30	23.68	17.14	13.43	11.04	9.37	8.14	7.20
		48.21	31.76	23.68	18.88	15.70	13.43	11.74
			55.38	38.30	29.27	23.68	19.89	17.14
Recip–	45.0			60.81	43.69	34.09	27.95	23.68
					65.05	48.21	38.30	31.76
						68.48	52.07	42.00
							71.29	55.38
								73.64
23.96	13.61	9.50	7.30	5.93	4.99	4.31	3.79	3.38
	38.66	23.96	17.36	13.61	11.19	9.50	8.26	7.30
		48.59	32.09	23.96	19.11	15.90	13.61	11.90
			55.76	38.66	29.58	23.96	20.13	17.36
Recip–	46.0			61.16	44.06	34.43	28.26	23.96
					65.39	48.59	38.66	32.09
						68.80	52.45	42.37
							71.60	55.76
								73.93
24.49	13.95	9.76	7.50	6.09	5.13	4.43	3.90	3.48
	39.35	24.49	17.78	13.95	11.48	9.76	8.48	7.50
		49.32	32.73	24.49	19.56	16.29	13.95	12.20
			56.47	39.35	30.19	24.49	20.60	17.78
Recip–	48.0			61.85	44.78	35.09	28.85	24.49
					66.05	49.32	39.35	32.73
						69.42	53.17	43.08
							72.18	56.47
								74.48

Table A1–1: (continued)

				Number of classes–					
2	3	4	5	6	7	8	9	10
25.00	14.28	10.00	7.69	6.25	5.26	4.55	4.00	3.57
	40.00	25.00	18.18	14.28	11.76	10.00	8.69	7.69
		50.00	33.33	25.00	20.00	16.67	14.28	12.50
			57.14	40.00	30.77	25.00	21.05	18.18
Recip–	50.0			62.49	45.46	35.71	29.42	25.00
					66.66	50.00	40.00	33.33
						70.00	53.85	43.75
							72.73	57.14
								75.00
26.19	15.07	10.58	8.15	6.63	5.58	4.82	4.25	3.79
	41.51	26.19	19.13	15.07	12.43	10.58	9.20	8.15
		51.56	34.74	26.19	21.01	17.55	15.07	13.20
			58.67	41.51	32.11	26.19	22.11	19.13
Recip–	55.0			63.95	47.01	37.16	30.73	26.19
					68.03	51.56	41.51	34.74
						71.30	55.40	45.29
							73.95	58.67
								76.15
27.27	15.79	11.11	8.57	6.98	5.88	5.08	4.48	4.00
	42.86	27.27	20.00	15.79	13.04	11.11	9.68	8.57
		52.94	36.00	27.27	21.95	18.37	15.79	13.85
			60.00	42.86	33.33	27.27	23.07	20.00
Recip–	60.0			65.21	48.39	38.46	31.92	27.27
					69.22	52.94	42.86	36.00
						72.41	56.76	46.67
							75.00	60.00
								77.14
28.26	16.45	11.61	8.97	7.30	6.16	5.33	4.69	4.19
	44.07	28.26	20.80	16.45	13.61	11.61	10.12	8.97
		54.17	37.14	28.26	22.80	19.12	16.45	14.44
			61.18	44.07	34.43	28.26	23.96	20.80
Recip–	65.0			66.32	49.62	39.63	33.00	28.26
					70.26	54.17	44.07	37.14
						73.39	57.96	47.89
							75.91	61.18
								78.00
29.17	17.07	12.07	9.33	7.61	6.42	5.56	4.89	4.37
	45.16	29.17	21.54	17.07	14.14	12.07	10.53	9.33
		55.26	38.18	29.17	23.59	19.81	17.07	15.00
			62.22	45.16	35.44	29.17	24.78	21.54
Recip–	70.0			67.30	50.73	40.70	33.98	29.17
					71.18	55.26	45.16	38.18
						74.24	59.04	49.00
							76.71	62.22
								78.75

Table A1–1: *(continued)*

	Number of classes–								
2	3	4	5	6	7	8	9	10
30.00	17.64	12.50	9.68	7.90	6.66	5.77	5.08	4.55
	46.16	30.00	22.22	17.64	14.63	12.50	10.91	9.68
		56.25	39.13	30.00	24.32	20.45	17.64	15.52
			63.16	46.16	36.36	30.00	25.53	22.22
Recip–	75.0			68.18	51.73	41.67	34.89	30.00
					71.99	56.25	46.16	39.13
						75.00	60.00	50.00
							77.42	63.16
								79.41
30.77	18.18	12.90	10.00	8.17	6.89	5.97	5.26	4.71
	47.06	30.77	22.86	18.18	15.09	12.90	11.27	10.00
		57.14	40.00	30.77	24.99	21.05	18.18	16.00
			64.00	47.06	37.21	30.77	26.23	22.86
Recip–	80.0			68.96	52.63	42.55	35.72	30.77
					72.72	57.14	47.06	40.00
						75.68	60.87	50.91
							78.05	64.00
								80.00
31.48	18.68	13.28	10.30	8.42	7.11	6.16	5.43	4.86
	47.89	31.48	23.45	18.68	15.52	13.28	11.60	10.30
		57.95	40.80	31.48	25.62	21.61	18.68	16.45
			64.76	47.89	37.99	31.48	26.87	23.45
Recip–	85.0			69.67	53.46	43.37	36.48	31.48
					73.37	57.95	47.89	40.80
						76.28	61.66	51.74
							78.61	64.76
								80.53
32.14	19.15	13.64	10.59	8.66	7.31	6.34	5.59	5.00
	48.65	32.14	24.00	19.15	15.93	13.64	11.92	10.59
		58.70	41.54	32.14	26.21	22.13	19.15	16.87
			65.45	48.65	38.71	32.14	27.48	24.00
Recip–	90.0			70.31	54.22	44.12	37.19	32.14
					73.97	58.70	48.65	41.54
						76.83	62.38	52.50
							79.12	65.45
								81.00
32.76	19.59	13.97	10.86	8.88	7.51	6.51	5.74	5.14
	49.35	32.76	24.52	19.59	16.31	13.97	12.22	10.86
		59.37	42.22	32.76	26.75	22.62	19.59	17.27
			66.09	49.35	39.38	32.76	28.04	24.52
Recip–	95.0			70.89	54.91	44.81	37.85	32.76
					74.50	59.37	49.35	42.22
						77.33	63.04	53.20
							79.58	66.09
								81.43

Table A1–1: *(continued)*

				Number of classes–				
2	3	4	5	6	7	8	9	10
33.33	20.00	14.29	11.11	9.09	7.69	6.67	5.88	5.26
	50.00	33.33	25.00	20.00	16.67	14.29	12.50	11.11
		60.00	42.86	33.33	27.27	23.08	20.00	17.65
			66.67	50.00	40.00	33.33	28.57	25.00
Recip–	100.0			71.42	55.56	45.45	38.47	33.33
					74.99	60.00	50.00	42.86
						77.78	63.64	53.85
							80.00	66.67
								81.82
34.37	20.75	14.86	11.58	9.48	8.03	6.96	6.14	5.50
	51.17	34.37	25.88	20.75	17.32	14.86	13.02	11.58
		61.11	44.00	34.37	28.20	23.91	20.75	18.33
			67.69	51.17	41.12	34.37	29.53	25.88
Recip–	110.0			72.36	56.70	46.61	39.57	34.37
					75.86	61.11	51.17	44.00
						78.57	64.71	55.00
							80.74	67.69
								82.50
35.29	21.43	15.38	12.00	9.84	8.33	7.23	6.38	5.71
	52.18	35.29	26.67	21.43	17.91	15.38	13.48	12.00
		62.07	45.00	35.29	29.03	24.66	21.43	18.95
			68.57	52.18	42.10	35.29	30.38	26.67
Recip–	120.0			73.17	57.69	47.62	40.54	35.29
					76.59	62.07	52.18	45.00
						79.25	65.63	56.00
							81.36	68.57
								83.08
36.11	22.03	15.85	12.38	10.16	8.61	7.47	6.60	5.91
	53.06	36.11	27.37	22.03	18.44	15.85	13.90	12.38
		62.90	45.88	36.11	29.76	25.32	22.03	19.50
			69.33	53.06	42.97	36.11	31.13	27.37
Recip–	130.0			73.86	58.56	48.51	41.41	36.11
					77.22	62.90	53.06	45.88
						79.82	66.43	56.87
							81.89	69.33
								83.57
36.84	22.58	16.28	12.73	10.45	8.86	7.69	6.80	6.09
	53.85	36.84	28.00	22.58	18.92	16.28	14.28	12.73
		63.64	46.67	36.84	30.43	25.93	22.58	20.00
			70.00	53.85	43.75	36.84	31.81	28.00
Recip–	140.0			74.46	59.32	49.30	42.17	36.84
					77.77	63.64	53.85	46.67
						80.33	67.13	57.65
							82.35	70.00
								84.00

Table A1–1: *(continued)*

				Number of classes–					
2	3	4	5	6	7	8	9	10	
37.50	23.07	16.67	13.04	10.72	9.09	7.89	6.98	6.25	
	54.55	37.50	28.57	23.07	19.35	16.67	14.63	13.04	
		64.29	47.37	37.50	31.03	26.47	23.07	20.45	
			70.59	54.55	44.44	37.50	32.43	28.57	
Recip–	150.0			75.00	60.00	50.00	42.86	37.50	
					78.25	64.29	54.55	47.37	
						80.77	67.74	58.33	
							82.76	70.59	
								84.37	
38.89	24.14	17.50	13.73	11.29	9.58	8.33	7.37	6.60	
	56.00	38.89	29.79	24.14	20.29	17.50	15.38	13.73	
		65.62	48.84	38.89	32.30	27.63	24.14	21.43	
			71.79	56.00	45.90	38.89	33.73	29.79	
Recip–	175.0			76.08	61.40	51.47	44.31	38.89	
					79.24	65.62	56.00	48.84	
						81.67	69.02	59.76	
							83.58	71.79	
								85.13	
40.00	25.00	18.18	14.29	11.77	10.00	8.70	7.69	6.90	
	57.15	40.00	30.77	25.00	21.05	18.18	16.00	14.29	
		66.67	50.00	40.00	33.33	28.57	25.00	22.22	
			72.73	57.15	47.06	40.00	34.78	30.77	
Recip–	200.0			76.92	62.50	52.63	45.46	40.00	
					79.99	66.67	57.15	50.00	
						82.35	70.00	60.87	
							84.21	72.73	
								85.71	
41.67	26.31	19.23	15.15	12.50	10.63	9.26	8.20	7.35	
	58.83	41.67	32.26	26.31	22.22	19.23	16.95	15.15	
		68.18	51.72	41.67	34.88	30.00	26.31	23.44	
			74.07	58.83	48.78	41.67	36.36	32.26	
Recip–	250.0			78.12	64.10	54.35	47.17	41.67	
					81.08	68.18	58.83	51.72	
						83.33	71.43	62.50	
							85.11	74.07	
								86.54	
42.86	27.27	20.00	15.79	13.05	11.11	9.68	8.57	7.69	
	60.00	42.86	33.33	27.27	23.08	20.00	17.65	15.79	
		69.23	52.94	42.86	35.99	31.03	27.27	24.32	
			75.00	60.00	50.00	42.86	37.50	33.33	
Recip–	300.0			78.94	65.22	55.56	48.39	42.86	
					81.81	69.23	60.00	52.94	
						84.00	72.42	63.64	
							85.72	75.00	
								87.10	

Table A1–1: *(continued)*

			Number of classes–						
2	3	4	5	6	7	8	9	10
44.44	28.57	21.05	16.67	13.80	11.76	10.26	9.09	8.16
	61.54	44.44	34.78	28.57	24.24	21.05	18.60	16.67
		70.59	54.55	44.44	37.49	32.43	28.57	25.53
			76.19	61.54	51.61	44.44	39.02	34.78
Recip–	400.0			80.00	66.67	57.14	50.00	44.44
					82.75	70.59	61.54	54.55
						84.85	73.69	65.12
							86.49	76.19
								87.80
45.45	29.41	21.74	17.24	14.29	12.19	10.64	9.43	8.47
	62.50	45.45	35.71	29.41	25.00	21.74	19.23	17.24
		71.43	55.56	45.45	38.45	33.33	29.41	26.32
			76.92	62.50	52.63	45.45	40.00	35.71
Recip–	500.0			80.64	67.57	58.14	51.02	45.45
					83.33	71.43	62.50	55.56
						85.37	74.47	66.04
							86.96	76.92
								88.23
46.15	30.00	22.22	17.65	14.64	12.49	10.91	9.68	8.70
	63.16	46.15	36.36	30.00	25.53	22.22	19.67	17.65
		72.00	56.25	46.15	39.12	33.96	30.00	26.87
			77.42	63.16	53.33	46.15	40.67	36.36
Recip–	600.0			81.08	68.18	58.82	51.73	46.15
					83.72	72.00	63.16	56.25
						85.71	75.00	66.67
							87.27	77.42
								88.52
47.37	31.03	23.08	18.37	15.26	13.04	11.39	10.11	9.09
	64.29	47.37	37.50	31.03	26.47	23.08	20.45	18.37
		72.97	57.45	47.37	40.29	35.06	31.03	27.83
			78.26	64.29	54.54	47.37	41.86	37.50
Recip–	900.0			81.81	69.23	60.00	52.95	47.37
					84.37	72.97	64.29	57.45
						86.30	75.91	67.74
							87.81	78.26
								89.01

APPENDIX 2
Classing by Specific Mathematical Curves

Numerous solutions to the problem of establishing unequal classes have been proposed. In discussing the "family of curves" concept under Reciprocal Curves, Chapter 10, we alluded to the possible use of specific types of mathematical curves established between the value extremes—such as arithmetic or logarithmic curves. A few words in regard to this approach may prove helpful.

If the value curve happened to be similar in its general trend to a logarithmic (or "geometric") curve drawn between the extreme values, it might be expected to serve as a classing curve. Logarithmic progressions are characterized by the fact that they increase at a constant rate as, for example, with such numbers as 2, 4, 8, and 16, or 40, 60, 90, and 135. Unfortunately, two difficulties are likely to be encountered when trying to use a logarithmic approach: First, whenever the minimum value happens to be zero—which is frequently the case—it is impossible to draw a logarithmic curve. Second, when a logarithmic curve *can* be drawn between the extreme values, it is rarely sufficiently similar in shape to the value curve to solve the problem. This second difficulty results from the fact that a logarithmic curve will vary greatly, depending on the particular minimum and maximum values between which it is drawn. For example, if the minimum value is a very small percentage of the maximum value, it will plot as an extremely low and sharply bent curve. At the other extreme, if the minimum value happens to be a very large percentage of the maximum value, it will plot as an almost straight line. Therefore, in any given instance, there is only a slim chance that the applicable logarithmic curve will be even roughly similar to the corresponding value curve—especially since even a small variation in the minimum value can cause a large change in the shape of the curve. In addition, a logarithmic curve can be used as a classing curve only when the value curve is low. Other types of mathematically defined curves (such as "arithmetic" and root curves) have somewhat corresponding limitations. Most importantly, all are specific curves.

What is really needed is a considerable "family of curves" from which to choose. Only through having a rather wide choice are we likely to find a curve appropriate to our needs—a simple curve that yields classes of increasing span and at the same time corresponds approximately to the general trend of the value curve.

By resorting to certain special mathematical manipulations, one may produce entire families of curves of various types, including the specific types mentioned. Most such curves, however, when plotted in a square such as we have been using for the display of the value curves, are asymmetrical about a diagonal line drawn between the upper left and lower right corners. Such asymmetry complicates matters by requiring the use of at least two families of curves, rather than one.

A mathematical curve that does not equalize, insofar as practical, the number of locations in each class while providing a smooth progression of class spans, is likely to be of little use. If the shape of the mathematical curve approximates the trend of the value curve (or is equally balanced in relation to it), it should be reasonably successful in meeting the desired goal. Before any mathematical curve is finally selected, the class breakpoints that it yields should be computed, and the number of locations falling in each class determined. As earlier suggested, however, the likelihood of a reasonably good fit in any specific instance is rather remote.

The reciprocal "family of curves" approach does not assure the best possible fit in every case, but for most value curves it does assure that a successful fit can be achieved easily, quickly, and consistently. For those who create thematic maps with any frequency, the achievement of consistent results in terms of a single type of classing curve may have particular value.

APPENDIX 3

Classing for Value Curves with Major Reversal

Value curves characterized by major reversals are those other than straight, low, or high types. Curves with one such reversal only will be the S or reverse-S type, but it is possible to have two major reversals (Figures A3–1A and A3–1B). When three or more reversals are present, they necessarily become smaller and hence should not be thought of as major reversals.

If a value curve with major reversal is more or less equally balanced in relation to a straight line curve, it is generally advisable to employ equal classing. However, there is at least one other possibility. If the curve is generally rather low or high and consequently has its point of reversal near one end, it is best to consider the reversals minor, and select an appropriate low or high reciprocal classing curve. With such value curves, any effort to have the classing system reflect the presence of the reversal would almost certainly prove confusing.

With intermediate situations, the reversal may be substantial, yet the curve as a whole may not be high or low enough to warrant use of a high or low classing curve. What is likely to be the best procedure?

Before seeking an answer to that question, let us give further consideration to the possible use of unequal classing for Curve 8 of Figure A3–1B; doing so will prove useful to what follows. Curve 8 happens to conform to the normal frequency distribution curve employed by statisticians. In order to prepare the drawing, it was necessary to assume that no values occur more than three standard deviations from the mean of all the values. (The standard deviation is the square root of the mean of the squared deviations from the base. In any normally distributed set of values some two-thirds [68.29 percent] of the values fall within one standard deviation of the mean, 95.45 percent fall within two standard deviations, and 99.73 percent fall within three standard deviations.)

At the top of Figure A3–2 is the normal frequency distribution curve as it is usually drawn. The lower-case Greek letter sigma (σ) is used in accor-

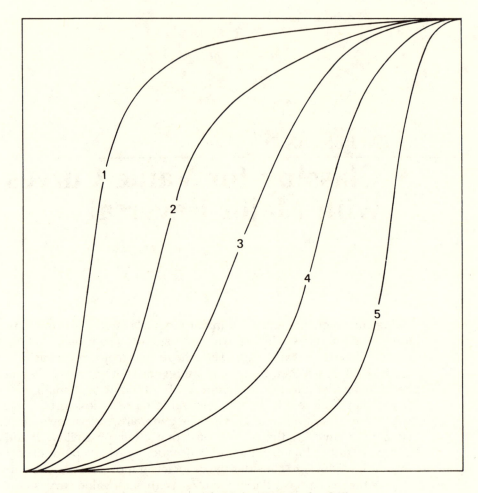

Figure A3–1A   Value Curves with Major Reversals, S Type

dance with established practice to represent "standard deviation." The curve is therefore three standard deviations below to three standard deviations above the mean. An infinity of possible values is assumed, and the height of the curve shows relative frequency per class, assuming a multiplicity of equal classes.

At the bottom of Figure A3–2, the normal curve has been transformed to show the directly corresponding value curve. It is a reverse-S type and corresponds to Curve 8 of Figure A3–1B. (The curve has here been represented by a continuous line.) As earlier stated, the range of values assumed is from minus 3σ through plus 3σ. If a greater range had been assumed, such as minus 5σ through plus 5σ, the center portion of the value curve would have been more nearly horizontal and the end portions more nearly vertical, with the bends far sharper in consequence. If the range had been smaller, such as minus 1.5σ through plus 1.5σ, the value curve would have been straighter overall.

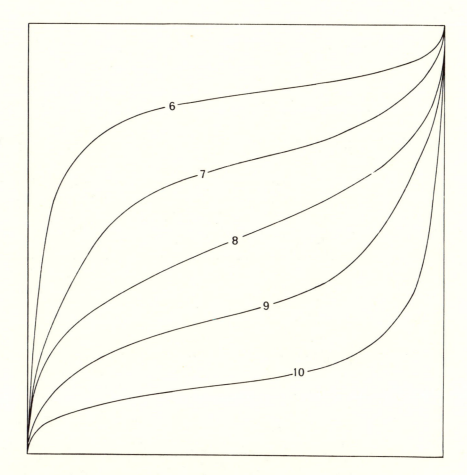

Figure A3–1B   Value Curves with Major Reversals, Reverse-S Type

Note that the point of reversal in the value curve occurs at the middle of the chart (designated by a small "x"). The point of curve reversal corresponds to the highest point of the normal frequency distribution curve shown above, which is the mean of all the values in the value set. (In the normal curve shown above, points of reversal occur at 1σ each side of the mean.)

Figure A3–3A shows how the assumed set of normally distributed values would appear if classed in 3, 4, 6, or 12 classes of equal span. In each case, the heights of the bars show the relative number of locations per class. For the series of diagrams on the left, the bars have been shown in contact, with their individual widths varied to reflect the class spans while keeping the overall chart widths constant. This permits their aggregate area in each case to be equal to the area under the normal curve at the top of Figure A3–2. For the series of diagrams on the right, the bars have been shown spaced apart with their widths constant. This has been done partly to give greater emphasis to the concept of "locations per class," but more particularly in anticipation of the need to represent unequal classing, for which the type

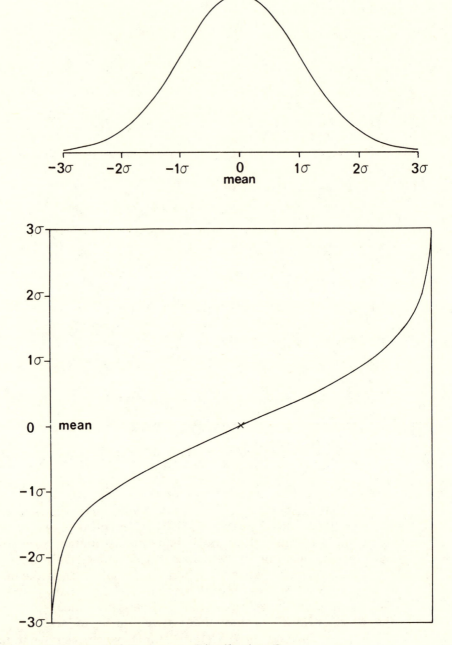

Figure A3–2   Normal Frequency Distribution Curve

of diagram on the left is troublesome because it cannot be used with any consistent vertical scale.

As suggested earlier, with this type of reverse-S value curve, the use of equal classing is the best way to bring out the nature of the distribution— to show that it is "normal" (high in the middle and low at each side). Figure

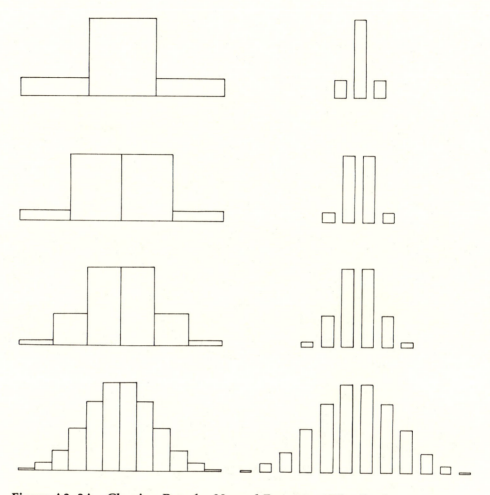

Figure A3–3A Classing Bars for Normal Frequency Distribution: 3, 4, 6, and 12 Classes

A3–3B presents a chart based on the use of six equal classes, and demonstrates the results achieved by means of the two sets of bars below.

If it were desirable to try to equalize the number of locations per class, the normal frequency curve itself might be used as a classing curve. In that event, in terms of our example and using five classes for variety (Figure A3–4), the value spans per class would become unequal, as shown by the lower bars. (Under the circumstances, the result would be the same as for quintile classing.)

Whenever a value curve and a classing curve agree, the number of locations per class will be equal. Hence, when a normal frequency curve is used as a classing curve, the degree to which the bars showing the number of locations per class are equal will be a measure of the degree to which the given values are symmetrically distributed. This feature justifies the occasional use of the normal frequency curve as a classing curve.

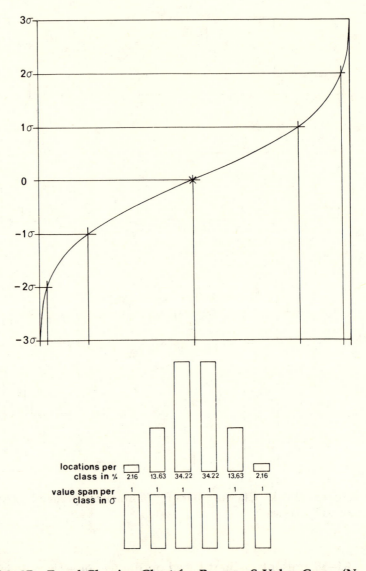

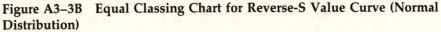

Figure A3–3B   Equal Classing Chart for Reverse-S Value Curve (Normal Distribution)

It should be noted that in diagrams of the type last shown, the heights of the upper bars correspond to the spacing of the projecting lines along the bottom of the chart, while the heights of the lower bars correspond to the spacing of the projecting lines along the left side of the chart. By putting the bars in proper alignment, good comparison can be made between the varying quantities involved.

As an alternative to the procedures just considered, a medial class might be used for all values within one standard deviation of the mean—with two classes on each side, each with a span of one standard deviation. The diagram would then appear as shown in Figure A3–5. As may be seen, in comparison

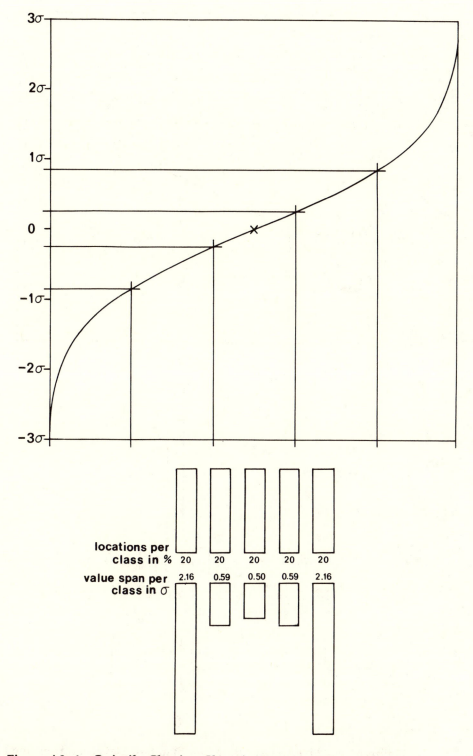

Figure A3–4 Quintile Classing Chart for Reverse-S Curve (Normal Distribution)

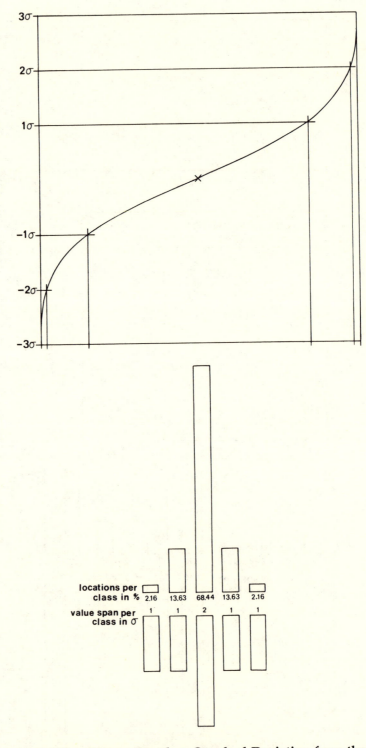

Figure A3–5 Five Classes Based on Standard Deviation from the Mean, Reverse-S Value Curve (Normal Distribution)

to the previous procedure shown in Figure A3–4, this greatly increases the span and hence the number of locations in the middle class.

Assuming the availability of suitable symbolism, six equal classes (as shown in Figure A3–3B)—three on each side of the mean and each equal to one standard deviation—are far better under most circumstances than the procedure just considered. This is the case not only because equal classing is preferable to unequal classing, but also because more information is provided and extreme differences in class span avoided. The use of six equal classes also shows relationships to the mean, which might be useful.

Let us return now to the problem of a value curve of reverse-S type and intermediate height, such as Curve 9 of Figure A3–1B. Assuming the use of six equal classes, the result would be as shown in Figure A3–6. Here, as is usual with equal classing, where the trend of the value curve is steep, the number of locations per class is small; and where it is flat, the number of locations per class is large.

Could unequal classing be used to advantage with such a curve? If the point of approximate curve reversal is used as the medial base, with a set of three classes of equal span on each side of that base, the result would be as shown in Figure A3–7. By this procedure, a substantially skewed distribution would be classed relative to the high point of the applicable frequency distribution curve. It is doubtful whether enough has been gained to justify this resort to unequal classing. Is there perhaps any other system of unequal classing that might prove more effective than equal classing?

The answer is no, except for the quantile system. Figure A3–8 shows the use of quantile in five classes. In Figure A3–6, the lower bars were of equal height, while now the upper bars are. The choice between the two methods depends upon the specific circumstances. Equal classing would be more usual, almost certainly easier to understand, and hence preferable under most circumstances. However, the quantile method tends to equalize the area of the map falling in each class, which is advantageous under some circumstances. With equal classing the emphasis is on the values involved in each class regardless of the number of locations. With the quantile method the emphasis is on the locations involved in each class regardless of the actual values.

For Curve 7 of Figure A3–1B, the problem is similar to that just discussed but in reverse. Once again, and for the same reasons, equal classing is probably preferable under most circumstances.

For value curves of the S type, as shown by Figure A3–1A, corresponding procedures might be used. Let us start by giving consideration to Curve 3, which is equally balanced in relation to a straight line curve. For this, as for Curve 8 of the reverse-S type, it would be advisable as a general rule to employ equal classing. With six equal classes, the result is that shown in Figure A3–9.

For Curves 1 and 5 of Figure A3–1A, it is best to consider the reversals minor, and to select an appropriate high or low reciprocal classing curve as previously recommended for Curves 6 and 10. For left- and right-weighted Curves 2 and 4, equal classing is desirable under most circumstances. Figures A3–10A and B show each of these curves classed in six equal classes.

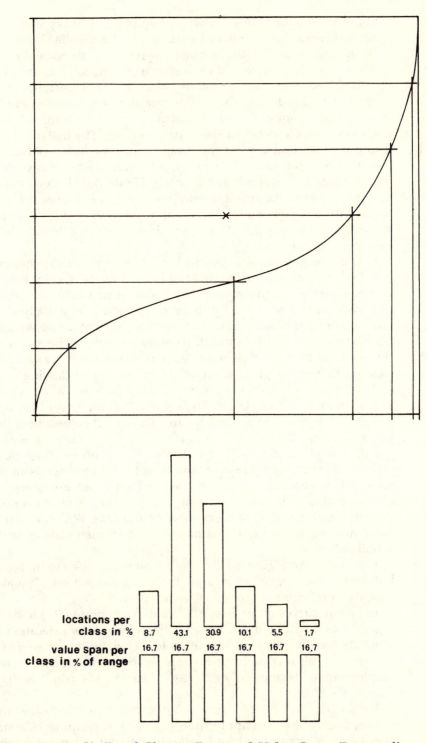

Figure A3–6   Six Equal Classes, Reverse-S Value Curve (Intermediate Height)

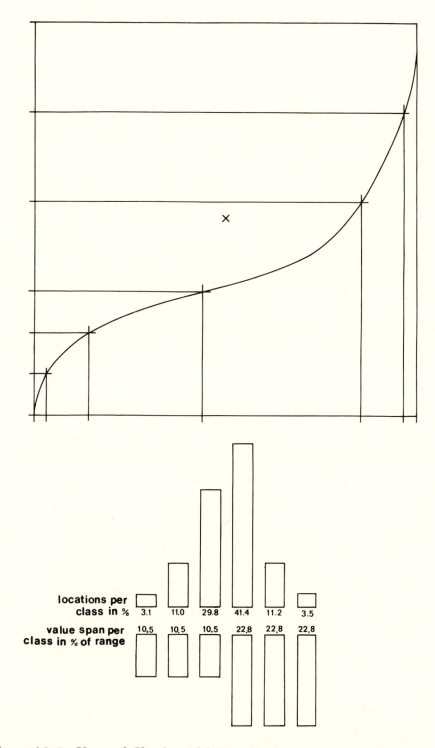

Figure A3–7   Unequal Classing with Two Groups of Three Equal Classes Around a Medial Point

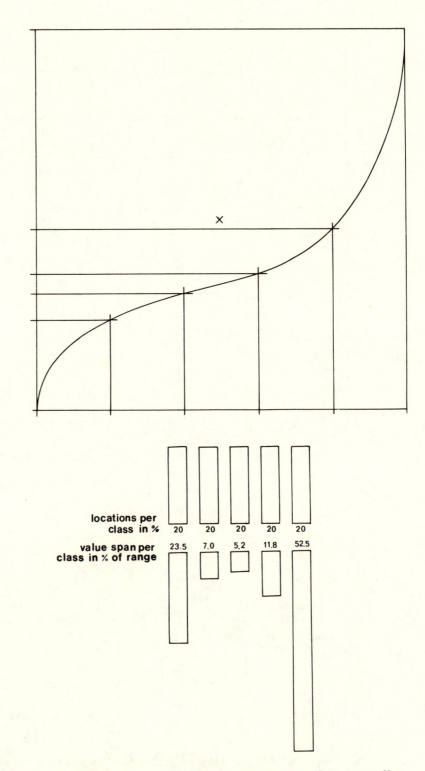

Figure A3–8   Quintile Classing, Reverse-S Value Curve (Intermediate Height)

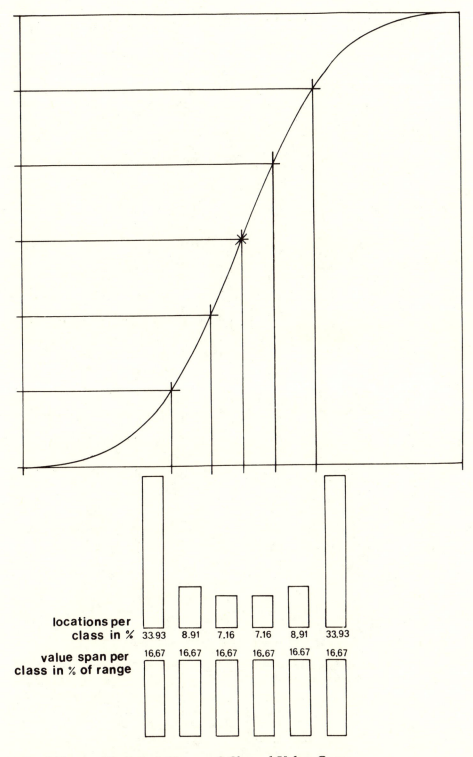

Figure A3–9   Six Equal Classes, S-Shaped Value Curve

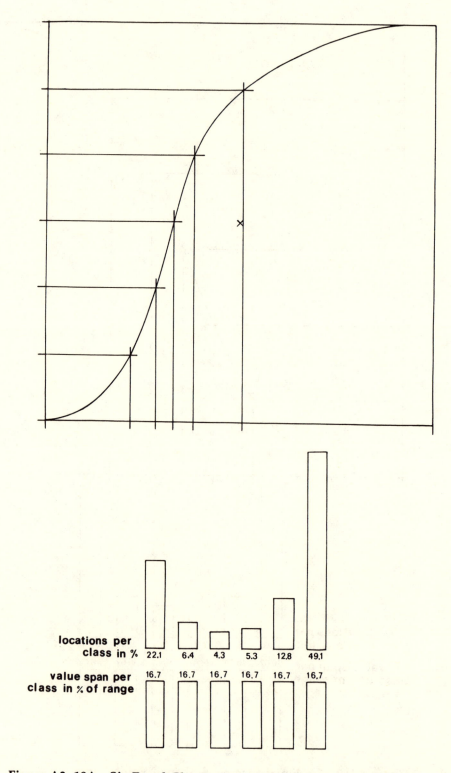

Figure A3–10A   Six Equal Classes, Left-Weighted S-Shaped Value Curve

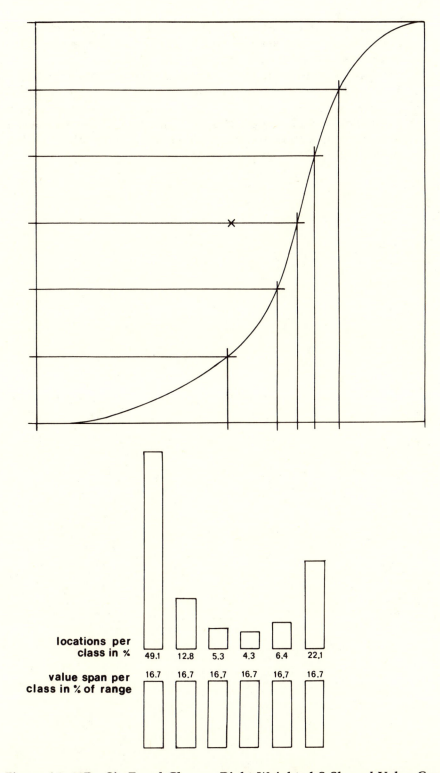

Figure A3–10B   Six Equal Classes, Right-Weighted S-Shaped Value Curve

Of all unequal classing procedures, quantile appears to be most valuable under typical circumstances. For Curves 2 and 4, the results of using quantile in six classes are shown in Figures A3–11A and B.

We need only briefly mention curves such as those shown in Figures A3–12A and B with two major reversals. Curves *b* and *e* are more or less equally balanced in relation to a straight line curve, and with them it would be advisable to employ equal classing. This would also probably be the case with the others shown, though for Curves *a* and *f*, high and low reciprocal curves, respectively, might be used.

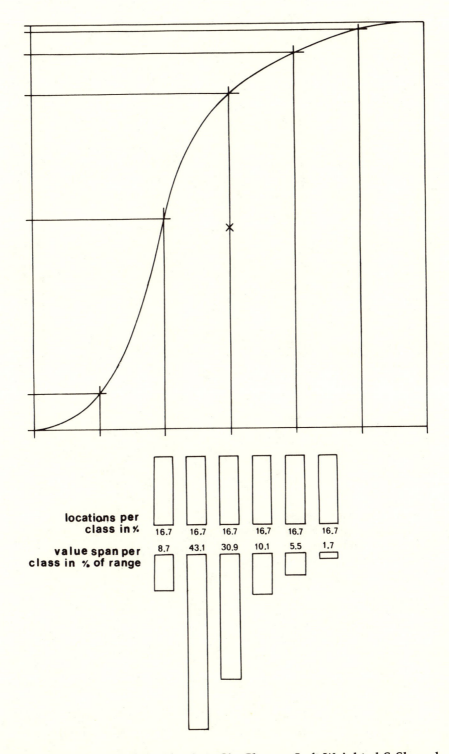

Figure A3–11A   Quantile Classing, Six Classes, Left-Weighted S-Shaped Value Curve

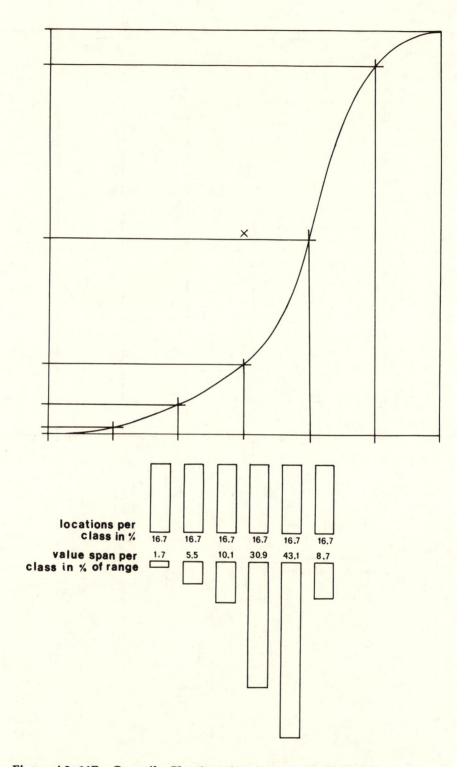

Figure A3–11B   Quantile Classing, Six Classes, Right-Weighted S-Shaped Value Curve

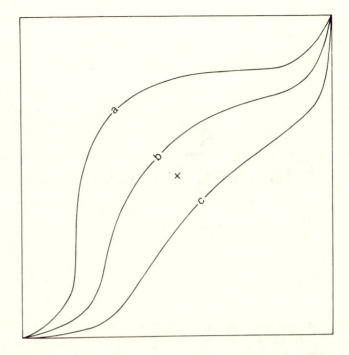

Figure A3–12A Value Curves with Two Major Reversals, High Curves

Figure A3–12B Value Curves with Two Major Reversals, Low Curves

APPENDIX 4

Classing Charts

A complete classing chart should show four things graphically: the value curve, the classing curve, the relative size of the class spans, and the relative number of locations falling in each class. (On the charts used to illustrate value curves with major reversal, classing curves were omitted for simplicity.) When space permits, the precise values delimiting each class span and the precise number of locations falling in each class may also be shown. However, in each chart, the nature of the classing curve should be emphasized. For simplicity in the following charts, the value curve has been represented by a continuous light line, and the classing curve has been drawn heavier. Hence, the value curve may be obscured in whole or in part by the classing curve; however, when this happens we may be sure that there is a good "fit" between the two.

Figure A4–1 shows a series of four such charts for equal classing, as applied to the original Foursquare values. In each case, the largely obscured value curve conforms to that of the upper diagram of Figure 10–3. As in other charts, the horizontal lines show the classing breakpoints, and the spacings determined by these lines, in combination with the top and bottom chart borders, show the relative sizes of the class spans. The spacing in each case is equal throughout, except for the slight inequality in Figure A4–1B, for three classes. (The class spans there were slightly rounded to 33, 34, and 33 in order to avoid the fractional values that would have resulted by using three precisely equal classes between 0 and 100 inclusive. With this exception, the classes for each of these diagrams are all of the same span.)

The small circles indicating the class breakpoints are shown centered at each intersection of the horizontal lines and the classing curve. The upper vertical lines also pass through the centers of these circles. When these lines are light, as in these four charts, they are equally spaced between the side borders. While not strictly necessary, they can frequently assist in drawing the classing curve or in revealing inaccuracy should they fail to meet the

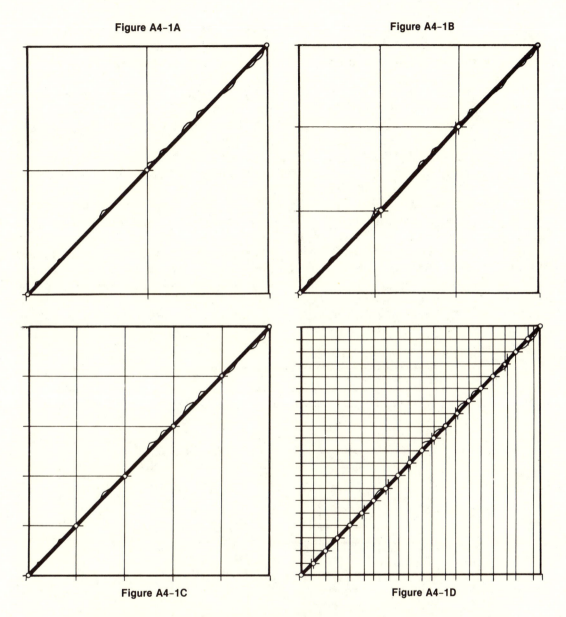

Figure A4–1A   Figure A4–1B

Figure A4–1C   Figure A4–1D

Figure A4–1   Classing Charts for Foursquare Value Set: 2, 3, 5, and 20 Classes

horizontal lines and classing curve line at the centers of the small circles. If equal classing is employed, the classing curve will always be precisely straight, but, as we shall see, it may also be straight under certain other circumstances. In Figure A4–1B, the classing curve departs slightly from a straight line because of the slightly unequal classing.

The positioning of the lower vertical lines, which serves to reveal the number of locations in each class, is determined by where the horizontal lines meet the value curve. A slight adjustment may be required left or right to assure that they fall midway between pairs of dots. The horizontal lines, the vertical lines, and the value curve lines do not necessarily meet at a point.

In Figures A4–1A and C, the spacings of the lines projecting below the bottom border are equal—signifying that the number of locations falling in each class is equal. In Figure A4–1A, the number in each class happens to be 50. In Figure A4–1C, it happens to be 20.

Due to the slightly unequal classing used and the nature of the particular values present in Figure A4–1B, the number of locations per class varies— with 31 in the first class, 34 in the second class, and 35 in the third class. The spacing between the lines projecting below the bottom border is proportional to those numbers. Variation in the number of locations per class is also found in Figure A4–1D, for 20 classes. In this case the variation in the number of locations per class ranges from a minimum of 3 (in the last class) to a maximum of 7 (in the next-to-last class). As the classes are all of the same span, the variation here results exclusively from the nature of the particular values present.

The chart of Figure A4–2 also applies to the original Foursquare values,

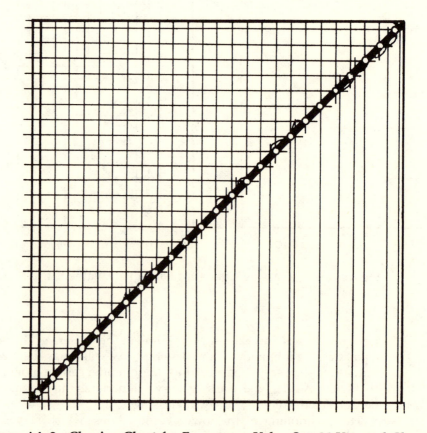

Figure A4–2 Classing Chart for Foursquare Value Set: 26 Unequal Classes

and again uses a precisely straight classing curve, producing 26 classes in this case. The classes, however, are unequal, the first and last being open-ended and only half as large as all the others. This is characteristic of dot maps (and maps of COUNT type) when class spans typically center on multiples of the rating per symbol. Note the use of heavier upper vertical lines adjacent to the left and right borders, between which the lighter lines are equally spaced. The number of locations per class varies from 1 to 7.

Let us turn now to the classing problem for the low value curve of Alternatives 2A and 2B. The chart of Figure A4–3 shows the use of five equal classes. Because of the nature of the value curve, the number of locations per class varies greatly, as may be seen by the spacing across the bottom. The actual number of locations per class is 58, 29, 10, 2, and 1. With such a low value curve and equal classing, it is characteristic to have many locations in the first class and only a few or even only one in the last class.

The use of equal classing with equal subclassing is illustrated in Figure A4–4. Figure A4–4A shows five equal classes with the first of the five divided into two equal subclasses (dashed lines). This procedure reduces the maximum number of locations per class from 58 to 34. Figure A4–4B shows five equal classes, the first two each divided into four equal subclasses. Conse-

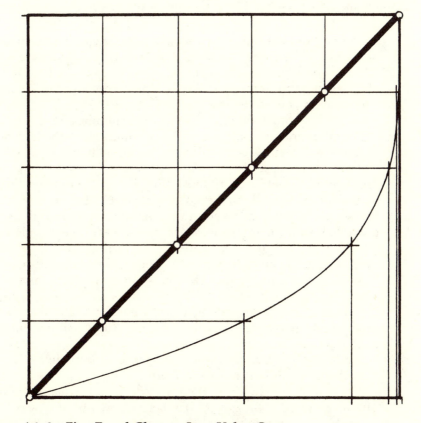

Figure A4–3   Five Equal Classes, Low Value Curve

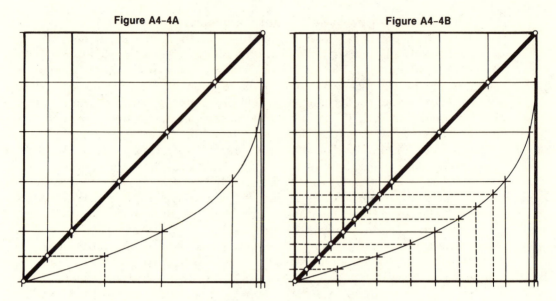

Figure A4–4A   Figure A4–4B

Figure A4–4   Five Equal Classes, Equal Subclassing

quently, the maximum number of locations per class is now reduced from 34 to 18.

Figure A4–5 again shows a straight classing curve with 26 classes. Because of the low value curve, the number of locations per class varies greatly, from 0 to 14. The first class has 7 locations compared to 14 for the second class, after which the number per class drops gradually until 7 empty classes are encountered (noted on the chart by "7 of 0"). Here again the final class contains only one location, the minimum number for a first or last class.

For other unequal classing methods applicable to the low value curve, see Figure A4–6. Figure A4–6A illustrates the quantile approach in terms of five classes. In this case the classing curve is polygonal, composed of straight line segments that result from placing the same number of locations (20) in each class. If the total number of locations (100) had not been evenly divisible by five, or if there had been more than one location of the same value at a quantile break, the number of locations in each class would have varied, in which event both the upper and lower vertical lines would have been unequally spaced. (This is why the upper lines have been drawn heavier, though in fact equally spaced.) Due to the orderly nature of the assumed low value curve, the classes are of progressively greater span, as shown by the spacing of the lines along the left border. Class spans are likely to be erratic with this system, however, due to the small irregularities that tend to characterize most value curves.

Figure A4–6B illustrates the use of a reciprocal curve. It corresponds to that illustrated at larger scale in Figure 10–16A.

In Figure A4–6C we see the effect of classing relative to a medial base. The mean of all values (20.6) has been shown by the heavier vertical line, on each side of which the lighter lines are equally spaced. The result is three

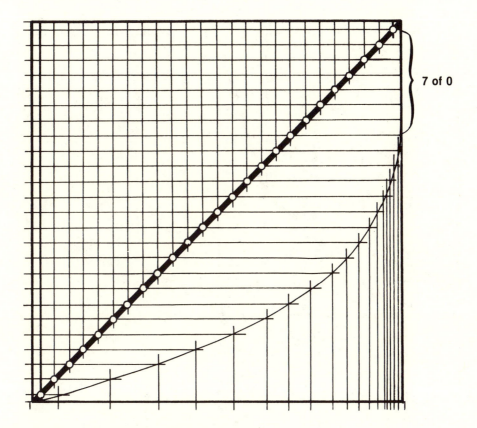

7 of 0

Figure A4–5   Twenty-six Equal Classes

small equal classes below the mean and three large equal classes above the mean. The number of locations per class varies from 1 to 33 (as compared to 13 to 29 for the preceding chart based on reciprocal classing).

Figure A4–6D shows the use of open-ended classing at both ends of the value range, with three equal classes between. Heavier vertical lines mark the rounded values (5 and 35) that were selected to serve as the basis for the open-ended classing. The lighter lines have been equally spaced between these.

The classing curve is dashed in part to stress the fact that the class spans of the two open-ended classes are unspecified. Such a procedure would have been used with Figures A4–2 and A4–5, except that space did not permit. Here, and in other charts based on open-ended classing, the small circles normally appearing at the ends of the classing curves have been omitted in further recognition of their being open-ended.

We will now turn to a brief consideration of the classing problem for the high value curve of Alternatives 3A and 3B. The chart of Figure A4–7 shows the use of five equal classes, as mapped in Figure 10–8B. As may be seen, the class spans are the opposite of those used for the corresponding low value curve of Figure A4–3. (This would not have been the case except for

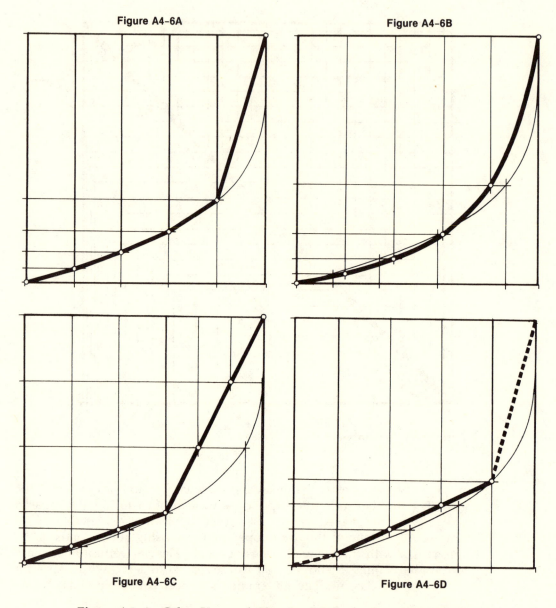

Figure A4–6A

Figure A4–6B

Figure A4–6C

Figure A4–6D

Figure A4–6   Other Unequal Classing Methods, Low Value Curve

the slight rounding that was employed.) However, because of the particular values present, the number of locations per class is slightly different (compare the value keys of Figures 10–7B and 10–8B). Any of the other classing methods used with low value curves might have been equally well employed.

The charting method illustrated here can be used to analyze any set of values. The preparation of such diagrams is useful if the map designer is in doubt as to the best classing method to employ, particularly if the value curve is characterized by significant irregularities or reversals. For example,

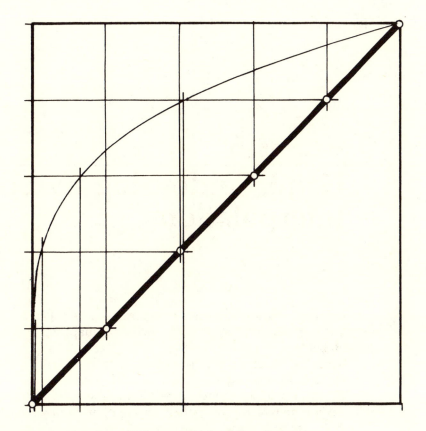

Figure A4–7   Five Equal Classes, High Value Curve

problems of the kind frequently encountered with the quantile method may be solved through the use of such diagrams made large enough to permit representation of each location by a separate dot, particularly if the equality that is characteristic of the quantile method is to be based on area or population.

Hand Contouring by Linear Interpolation

To construct a hand-contoured map, proceed as follows. (The first three steps here are identical to those for proximal mapping; see Appendix 7 for more detail.)

1. Mark the center of gravity of each data zone with a dot and number the dots for identification.
2. Establish proximal zones in relation to the dots.
3. Determine which proximal zones touch, and draw the lines connecting their centers.
4. Decide the number and range of the class intervals to be employed. (This step determines the values of the contour lines.)
5. Along the center lines, establish the points where contour lines will cross. Use the "Hand Contouring Form" (Appendix 6) to speed computations. (See Figure A5–1.)
6. Connect the points thus established with straight lines. Where contours meet an outermost center line, carry through to the limits of the study area by proceeding at right angles to the outermost center line. (If any two of the outermost center lines form a reentrant angle, proceed at right angles to a line connecting across the reentrant angle. See Figure A5–2.)
7. On tracing paper, draw freehand curved contours based on the contours thus drawn—departing from the original lines only enough to avoid angularity while still crossing the center lines at the points established under Item 5 above. (Omit all construction or other lines except the outline of the study area and the data points. See A5–3 and A5–4.)
8. Indicate by suitable graphic symbolism the relative elevations. Symbolism 9–FC8 will be the easiest to produce by hand, but should not be used with more than six classes (five contours). Symbolism 8–FE7 will be the most effective under typical circumstances, preferably with not more than eight classes. (See Figure A5–5 and A5–6.) These symbolisms are discussed in detail in Chapter 9.

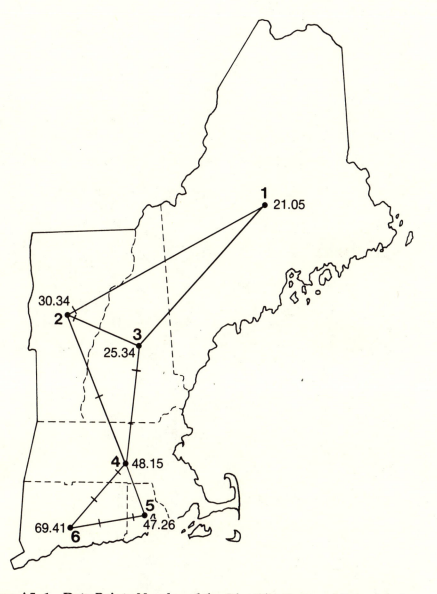

Figure A5–1 Data Points Numbered for Identification, and Interpolation Points Shown Where Contour Lines Will Cross Center Lines

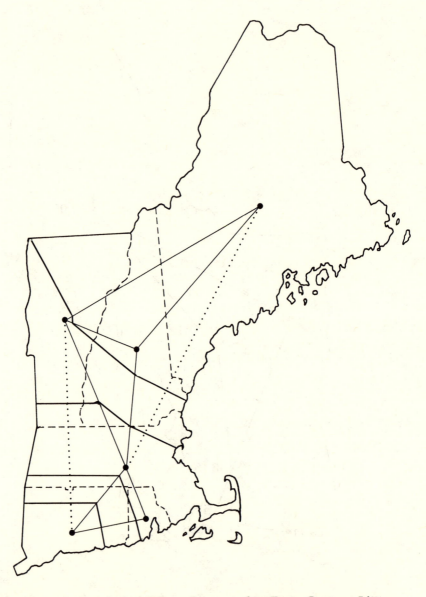

Figure A5–2   Interpolation Points Connected to Form Contour Lines

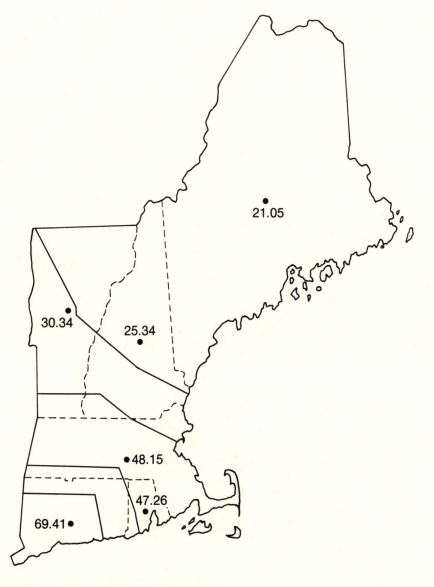

Figure A5–3 Contour Lines Stressed

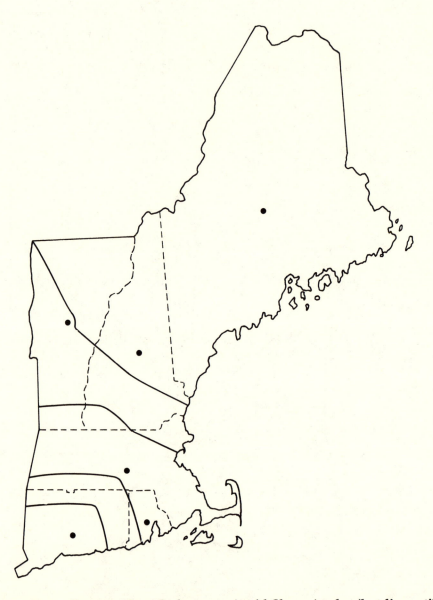

Figure A5–4   Contour Lines Redrawn to Avoid Sharp Angles (but lines still passing through interpolation points)

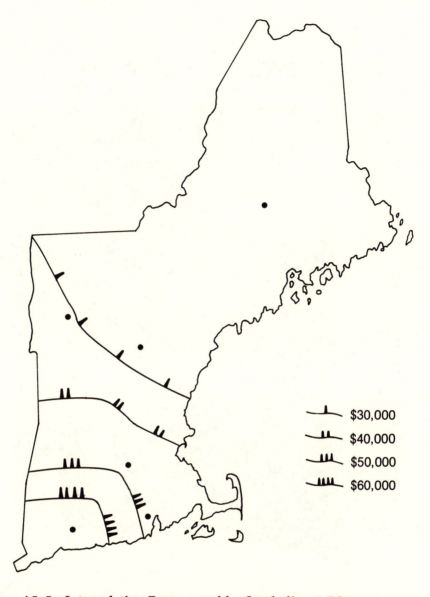

Figure A5–5   Interpolation Represented by Symbolism 9-FC8

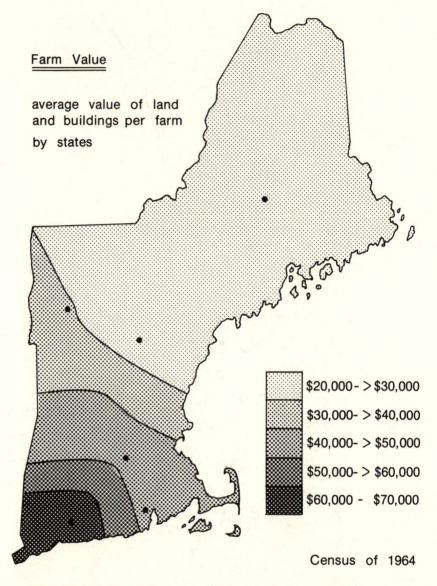

Figure A5–6   Interpolation Represented by Symbolism 8-FT7

Hand Contouring Form

One form such as this one would be used for each center line. All the figures to be placed in the boxes should be shown on a single source map, and then all the boxes for each sheet should be filled before starting work on that sheet. Those computations involving center distance (length in inches between points) should be performed last, after everything else on each sheet has been completed. (The data point pairs to be used in determining center distance are those whose proximal zones meet within the study area at more than one point.)

This sheet is for the center line that connects points 4 and 3

Higher point value (point 4)	= [48.15]
Lower point value (point 3)	= [25.34]
Reciprocal of the difference	1/(22.81) = (.044)

	Length in inches bet. 2 val. pts.	Compute last

First contour

Value above lower point value = [30.00]
Subtract lower point value = [25.34]

(4.66) × (.044) = (.205) × [1.24] = (.25)

Second contour

Value above lower point value = [40.00]
Subtract lower point value = [25.34]

(14.66) × (.044) = (.645) × [1.24] = (.80)

Third contour

Value above lower point value = [.]
Subtract lower point value = [.]

(.) × (.044) = (.   ) × [1.24] = (.)

	Length in inches bet. 2 val. pts.	Compute last

Fourth contour
Value above lower point value   = [.]
Subtract lower point value   = [.]

(.) × (.044) = (.   ) × [1.24] = (.)

Fifth contour
Value above lower point value   = [.]
Subtract lower point value   = [.]

(.) × (.044) = (.   ) × [1.24] = (.)

Sixth contour
Value above lower point value   = [.]
Subtract lower point value   = [.]

(.) × (.044) = (.   ) × [1.24] = (.)

Seventh contour
Value above lower point value   = [.]
Subtract lower point value   = [.]

(.) × (.044) = (.   ) × [1.24] = (.)

Eighth contour
Value above lower point value   = [.]
Subtract lower point value   = [.]

(.) × (.044) = (.   ) × [1.24] = (.)

Ninth contour
Value above lower point value   = [.]
Subtract lower point value   = [.]

(.) × (.044) = (.   ) × [1.24] = (.)

APPENDIX 7

Constructing Proximal Maps

Proximal maps are constructed as follows:
1. Indicate base centers of locations (or approximate centers of gravity) by dots. Draw centerlines connecting them, so as to form a network of triangles within the study area. (Insofar as convenient, try to form acute rather than obtuse triangles. See Figure A7–1.)
2. Draw perpendicular bisectors across the midpoint of each line, continuing each bisector far enough to cross the bisectors for the other two sides of each triangle or the limit of the study area. (See Figure A7–2.)
3. Retrace more heavily the outlines of the proximal zones thus created. (See Figure A7–3.)
4. Add tone to show the proximal deviation—the difference between the original zones and the newly created ones. (In the illustrated map, the southern tip of Maine now has the same value as most of New Hampshire. See Figure A7–4.)

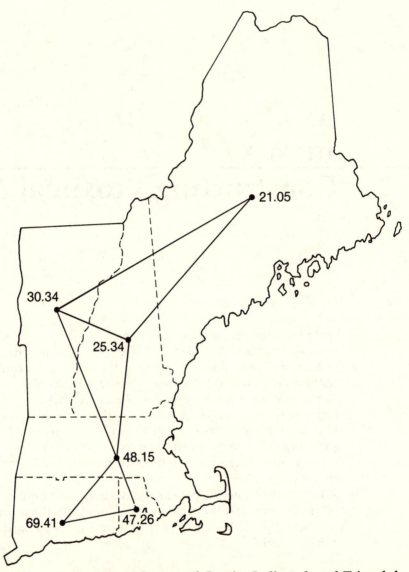

Figure A7–1   Approximate Centers of Gravity Indicated, and Trianglular Network of "Center Lines" Constructed (shallow obtuse triangles facing outward have been omitted, as recommended)

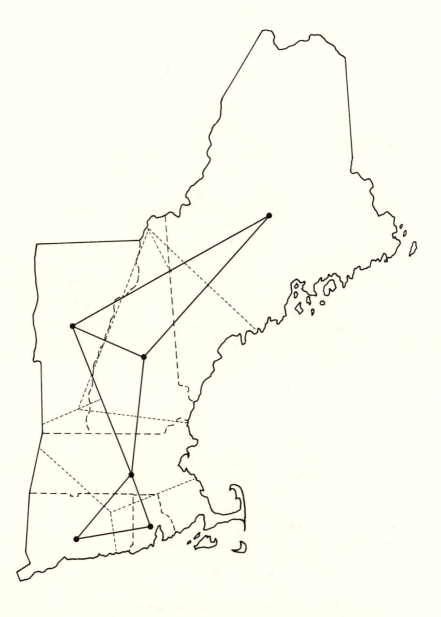

Figure A7–2   Perpendicular Bisectors of Center Lines Constructed

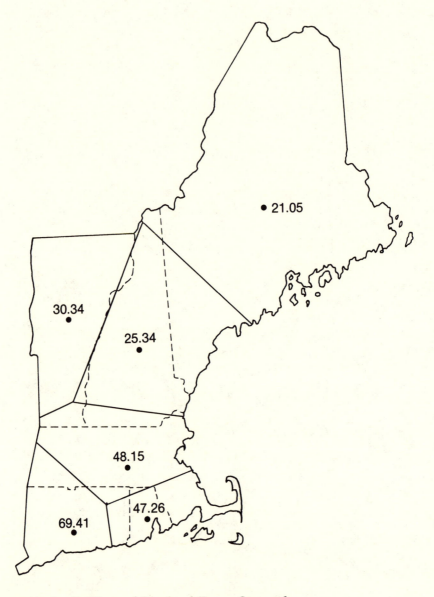

Figure A7–3   Outlines of Proximal Zones Stressed

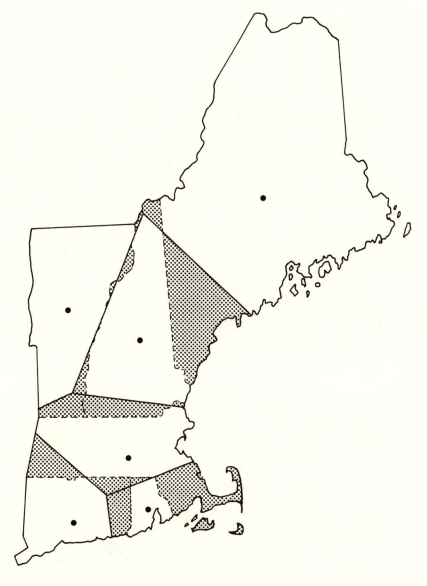

Figure A7–4   Proximal Deviation Shown by Tone

Traditional Dot Mapping by Hand

For the production of dasymetric dot maps of traditional type, it is necessary to employ the somewhat complex procedure outlined below—though in practice the various steps may be elided.

STEP 1

Start with given locations and their relevant values. The values will determine, in *Step 3*, the number of dots to be placed within each location base outline.

STEP 2

Any special knowledge regarding the spatial variability of the data within each location, or factors likely to influence such variability, should be compiled. Library, field, or other research should then be undertaken, to the extent warranted and practical. A particular effort should be made to achieve a consistent level of detail over the study space as a whole.

STEP 3

Consider suitable dot size and dot rating. The two goals are: densely spaced dots (about to coalesce) in those areas of the map where the highest values exist; and rather widely spaced dots in those areas of the map where the lowest values exist. Obviously, in order to make these decisions, the size of the map must also be established.

Frequently the goals previously stated will be impossible to achieve simultaneously. The mapmaker will be faced with having to choose between two undesirable alternatives: having large areas of the map entirely or largely devoid of dots, or using such small dots that they will be visually ineffective and difficult to produce or reproduce. The solution lies in establishing a sufficiently small dot rating to avoid having those areas of the map with low values largely devoid of dots, while using reasonably sized dots and exaggerating as necessary the extent of the high value areas. To minimize the need to exaggerate the high value areas, the dot rating should be no smaller than necessary.

Usually at least two or three trials will be made before an optimum approach for the particular circumstances is achieved. Such trials will involve matters to be considered below.

For any assumed dot rating, the number of dots to be used within a location will be determined by dividing the value for the location by the dot rating. If the result is a fraction, which is normally the case, the number of dots will depend on the classing system to be used. As mentioned previously, the most commonly employed procedure and the easiest to comprehend involves the establishment of class breaks at the midpoints between multiples of the dot rating to be used. For example, assuming a dot rating of 1000, three dots would represent the class span of 2500–<3500.

STEP 4

In terms of the map size, dot size, and dot rating under *Step 3*, the positioning of the dots applicable to high value areas of the map must next be determined. This will involve the use of special knowledge, and requires judgment. The best procedure will be to place the dots so that they almost coalesce. If the map is to be reproduced, consideration must also be given to the likelihood that as a consequence of reproduction processes, where the dots are almost coalescent, they tend to merge together into indistinguishable clusters. If the map is to be reproduced at a size other than the original, further allowance must be made in determining the spacing in the original so that the criteria suggested here are met in the map as it finally appears to the user.

The closely spaced dots in the high value areas of the map will almost always require more space than exists in those areas. Thus, as was indicated earlier, the size of the high value areas must be exaggerated. Clusters of densely spaced dots should be centered on and conform in general shape to the high value area represented. However, when a high value area lies wholly within one location, the exaggerated area should be kept wholly within the same location, even if its shape must be modified to make that possible. Only thus will the correct number of dots appear within each location.

STEP 5

After the positions of all densely spaced dots applicable to high value areas are established, the positions of all remaining dots may be determined. For this purpose, the use of special knowledge and judgment will again be mandatory. The final quality of the map is dependent primarily on the quality of the special knowledge that is *actually employed*.

Bibliography

Editor's note: This bibliography, with the exception of Section VII, is based on a compilation by Carolyn C. Weiss, formerly of the Laboratory for Computer Graphics and Spatial Analysis and now with Statistics Canada.

I. Books (general)

Bachi, R. 1968. *Graphical rational patterns: a new approach to graphical presentation of statistics.* New York: Israel Universities Press.

Bertin, J. 1967. *Semiologie graphique.* Paris: Gauthier-Villars.

Birch, T. W. 1964. *Maps topographical and statistical.* 2d ed. New York: Oxford Univ. Press.

Bowman, W. J. 1968. *Graphic communication.* New York: John Wiley.

Bunge, W. 1966. *Theoretical geography.* Lund, Sweden: Lund Series in Geography, C.W.K. Gleerup Publishers. (See Chapter 2, "Metacartography," pp. 38–71.)

Cole, J. P., and King, C. A. M. 1968. *Quantitative geography.* New York: John Wiley.

Davis, J. C., and McCullagh, M. V. 1975. *Display and analysis of spatial data.* New York: John Wiley.

Dickinson, G. C. 1973. *Statistical mapping and the presentation of statistics.* 2d ed. New York: Crane, Russak & Co.

Experimental Cartography Unit, Royal College of Art. 1971. *Automatic cartography and planning.* London: Architectural Press.

Gould, P., and White, R. 1974. *Mental maps.* New York: Pica Press.

Greenhood, D. 1964. *Mapping.* Chicago: Univ. of Chicago Press.

Hsu, M. L., and Robinson, A. H. 1970. *The fidelity of isopleth maps: an experimental study.* Minneapolis: Univ. of Minnesota Press.

Keates, J. S. 1973. *Cartographic design and production.* New York: John Wiley.

Laboratory for Computer Graphics and Spatial Analysis. 1966–1972. *Selected projects.* Cambridge: Harvard University.

Lawrence, G. R. P. 1971. *Cartographic methods.* London.

Monkhouse, F. J., and Wilkinson, H. R. 1971. *Maps and diagrams.* 3d ed. New York: E. P. Dutton.

Muehrcke, P. 1978. *Map use*. Madison, Wis.: J. P. Publications.

Mulvey, F. 1969. *Graphic perception of space*. New York: Reinhold.

Raisz, E. 1962. *Principles of cartography*. New York: McGraw-Hill.

Robinson, A. 1952. *The look of maps*. Madison, Wis.: Univ. of Wisconsin Press.

————. 1981. *Early thematic mapping in the history of cartography*. Chicago: Univ. of Chicago Press.

Robinson, A. H., and Petchenik, B. B. 1976. *The nature of maps: essays toward an understanding of maps and mapping*. Chicago: Univ. of Chicago Press.

Robinson, A. H., and Sale, R. D. 1976. *Elements of cartography*. 4th ed. New York: John Wiley.

II. Monographs (general)

American Association of Petroleum Geologists. 1970. *Slide manual: a guide to the preparation and use of projection slides*. 3d ed.

Castner, H. W., and McGrath, G., eds. 1971. *Map design and the map user. Cartographica*, Monograph no. 2.

Castner, H. W., and Robinson, A. H. 1969. *Dot area symbols in cartography: the influence of pattern on their perception*. ACSM Monographs in Cartography, no. 1. American Congress on Surveying and Mapping. Washington, D.C.

Kingsbury, R. C. 1969. *Creative cartography: an introduction to effective thematic map design*. Occasional Publication no. 4. Department of Geography, Indiana University.

Moore, L. C. 1970. *Cartographic scribing materials, instruments, and techniques*. Technical Monograph no. CA–3. Cartography Division, American Congress on Surveying and Mapping. Washington, D.C.

Morrison, J. L. 1971. *Method-produced error in isarithmic mapping*. Technical Monograph no. CA–5. Cartography Division, American Congress on Surveying and Mapping. Washington, D.C.

Muehrcke, P. 1972. *Thematic cartography*. Commission on College Geography, Resource Paper no. 19. Association of American Geographers.

Peucker, T. K. 1972. *Computer cartography*. Commission on College Geography, Resource Paper no. 17. Association of American Geographers.

Steward, H. J. 1974. *Cartographic generalization: some concepts and explanation. Cartographica*, Monograph no. 10.

Thomas, E. N. 1960. *Maps of residuals from regression: their characteristics and uses in geographic research*. Geography Publication no. 2, University of Iowa. (Also appears in *Spatial analysis: a reader in statistical geography*, 1968, ed. J. L. Berry and D. F. Marble, Englewood Cliffs, N.J.: Prentice-Hall. Pp. 326–52.)

Williams, R. L. 1956. *Statistical symbols for maps: their design and relative values*. Map Laboratory, Yale University.

III. Conformant (Choropleth) Mapping

Dixon, O. M. 1972. Methods and progress in choropleth mapping of population density. *Cartographic Journal* 9:19–29.

Monmonier, M. S. 1975. Pre-aggregation of small areal units: a method of improving communication in statistical mapping. *Proceedings of the American Congress on Surveying and Mapping*, 35th Annual Meeting, pp. 260–269.

IV. Contour (Isarithmic) Mapping

Barnes, J. A. 1956. Some basic problems concerned with isoline mapping. Paper read at the Georgia Academy of Science, Annual Meeting, Atlanta, Georgia.

————. 1957. Control areas and control points in isopleth mapping. Paper read at the American Association of Geographers' Annual Meeting, Cincinnati, Ohio. Available from the Laboratory for Computer Graphics and Spatial Analysis, Harvard University.

Blumstock, D. I. 1953. The reliability factor in the drawing of isarithms. *Annals, Association of American Geographers*, vol. 43, no. 4.

Clark, P. J., and Evans, F. C. 1954. Distance to nearest neighbor as a measure of spatial relationships in populations. *Ecology*, vol. 35, no. 4.

Hsu, M. L. 1968. The isopleth surface in relation to the system of data derivation. *International yearbook of cartography*, vol. 8.

Hsu, S. Y., and Lau, J. 1974. Determining sub-optimal solutions of grid cell distance and interpolation schemes in isopleth mapping. *Proceedings of the American Congress on Surveying and Mapping*, Fall Convention.

Kolberg, D. W. 1974. Population aggregations as a continuous surface: an example of computer mapping. *Cartographic Journal*, vol. 11, no. 2.

Morrison, J. L. 1969. Control point spacing as an indicator of the accuracy of isarithmic maps. *Proceedings of the American Congress on Surveying and Mapping*, 29th Annual Meeting.

————. 1970. A link between cartographic theory and mapping practice: the nearest neighbor statistic. *Geographic review*, vol. 60, no. 4.

————. 1974. Observed statistical trends in various interpolation algorithms useful for first stage interpolation. *Canadian Cartographer*, vol. 11, no. 2.

Schmid, C. F., and MacCannell, E. H. 1955. Basic problems, techniques and theory of isopleth mapping. *Journal of the American Statistical Association*, vol. 50, no. 269.

Shepard, D. S. 1968. A two dimensional interpolation function for computer mapping of irregularly spaced data. *Harvard papers in theoretical geography*, Geography and the Properties of Surfaces, no. 15.

————. 1970. SYMAP interpolation characteristics. *Computer mapping as an aid in air pollution studies*. Vol. 2: Individual Reports, Grant no. 68A–2405D. National Air Pollution Control Administration, Public Health Service, U.S. Department of Health, Education and Welfare.

V. Trend Surfaces and Residual Mapping

Chorley, R. J., and Haggett, P. 1965. Trend surface mapping in geographical research transactions and papers. *The Institute of British Geographers*, no. 37, pp. 47–67. Also appears in *Spatial analysis: a reader in statistical geography*, 1968, ed. B. J. L. Berry and D. F. Marble, chap. 7. Englewood, Cliffs, N.J.: Prentice-Hall.

Harbough, J. W., and Merriam, D. F. 1967. *Computer applications in stratigraphic analysis*, New York: John Wiley. Chap. 5: Polynomial trend analysis, pp. 61–87.

VI. Value Classing Procedures

Chang, K. T. 1974. An instructional computer program on statistical class intervals. *Canadian cartographer*, vol. 11, no. 1.

Jenks, G. F. 1963. Generalization in statistical mapping. *Annals, Association of American Geographers*, vol. 53, no. 1.

Jenks, G. F. and Caspall, F. C. 1971. Error on choropleth maps: definition, measurement, reduction. *Annals, Association of American Geographers*, vol. 61, no. 2.

Jenks, G. F. and Coulson, M. R. C. 1963. Class intervals for statistical maps. *International yearbook of cartography*, vol. 3.

Monmonier, M. S. 1972. Contiguity-biased class-interval selection: a method for simplifying patterns on statistical maps. *Geographical review*, vol. 62.

————. 1973. Analogs between class-interval selection and location-allocation models. *Canadian Cartographer*, vol. 10, no. 2.

Scripter, M. W. 1970. Nested-means map classes for statistical maps. *Annals, Association of American Geographers*, vol. 60, no. 2.

VII. MAP PRODUCTION AND REPRODUCTION

Bouma, D. G. 1962. The uses of photomechanical processes in cartography. Paper presented at 57th Annual Meeting of the Association of American Geographers, at Miami Beach, Florida.

Hodgkiss, A. G. 1970. *Maps for books and theses.* New York.

International Paper Co. 1979. *Pocket pal.* 12th ed. New York.

Keuffel & Esser Co. *Municipal map-making: color and economy with stabilene system materials.* Morristown, N.J.

Preucil, F. M. A new method of rating the efficiency of paper for color reproductions. Research Progress Report 60, General Memo no. 8. Graphic Arts Technical Foundation. Pittsburgh, Pa.

Shapiro, C., ed. 1968. *The lithographers manual.* Graphic Arts Technical Foundation. Pittsburgh, Pa.

VIII. OTHER ARTICLES

Blakemore, M. J., and Harley, J. B. 1981. Concepts in the history of cartography: a review and perspective. *Cartographia*, vol. 17, no. 4.

Peuker, T. K. 1972. Computer cartography: a working bibliography. University of Toronto, Department of Geography, Discussion Paper no. 12.

Index